AF360911

Formation of Advanced Nanomaterials by Gas-Phase Aggregation

Formation of Advanced Nanomaterials by Gas-Phase Aggregation

Editors

Vladimir N. Popok
Ondřej Kylián

MDPI • Basel • Beijing • Wuhan • Barcelona • Belgrade • Manchester • Tokyo • Cluj • Tianjin

Editors
Vladimir N. Popok
Aalborg University
Denmark

Ondřej Kylián
Charles University
Czech Republic

Editorial Office
MDPI
St. Alban-Anlage 66
4052 Basel, Switzerland

This is a reprint of articles from the Special Issue published online in the open access journal *Applied Nano* (ISSN 2673-3501) (available at: http://www.mdpi.com).

For citation purposes, cite each article independently as indicated on the article page online and as indicated below:

LastName, A.A.; LastName, B.B.; LastName, C.C. Article Title. *Journal Name* **Year**, *Volume Number*, Page Range.

ISBN 978-3-0365-2728-4 (Hbk)
ISBN 978-3-0365-2729-1 (PDF)

Cover image courtesy of Vladimir Popok and Ondřej Kylián

Contents

About the Editors

Vladimir N. Popok is a physicist by education. He has a PhD degree from Belarusian State University obtained in 1995. During his career, he worked at the University of Chemistry and Technology and Nuclear Physics Institute (both in Czechia), University of Gothenburg (Sweden) and Rostock University (Germany). Currently, he is an Associate Professor at the Department of Materials and Production of Aalborg University in Denmark. His main fields of research cover the modification of surfaces using cluster ion beams, formation of metal nanoparticles by gas aggregation towards applications for enhanced sensing and detection, development of metal–polymer nanocomposites, growth of III-N semiconductor materials and reliability of electronic components. He is author or co-author of over 130 articles in peer-reviewed journals, several book chapters and a number of proceedings' papers presented at international conferences/symposia including about 30 invited and plenary talks.

Onřej Kylián is a physicist and material scientist. After finishing his PhD studies in 2003 at the Charles University (Prague, Czech Republic), he was working for almost 5 years as a contractual agent at the Joint Research Centre of the European Commission (Ispra, Italy). Afterwards, he started to work as a Senior Assistant Professor and, more recently, as an Associate Professor at the Department of Macromolecular Physics of the Faculty of Mathematics and Physics, Charles University. Since 2020, he has acted as the head of this department. Currently, his main scientific interest is in the production and characterization of advanced functional nanomaterials by plasma-based techniques. He is author or co-author of more than 140 scientific papers published in peer-reviewed journals, 7 book chapters and more than 200 conference contributions including 13 invited and keynote lectures.

applied nano

Editorial

Formation of Advanced Nanomaterials by Gas-Phase Aggregation

Vladimir N. Popok [1,*] and Ondřej Kylián [2]

1   Department of Materials and Production, Aalborg University, 9220 Aalborg, Denmark
2   Department of Macromolecular Physics, Faculty of Mathematics and Physics, Charles University, 180 00 Prague, Czech Republic; ondrej.kylian@gmail.com
*   Correspondence: vp@mp.aau.dk; Tel.: +45-9940-9229

Citation: Popok, V.N.; Kylián, O. Formation of Advanced Nanomaterials by Gas-Phase Aggregation. *Appl. Nano* **2021**, 2, 82–84. https://doi.org/10.3390/applnano2010007

Received: 24 February 2021
Accepted: 16 March 2021
Published: 19 March 2021

Publisher's Note: MDPI stays neutral with regard to jurisdictional claims in published maps and institutional affiliations.

Gas aggregation is a well-known phenomenon, often seen in nature under temperature lowering, as, for example, cloud, fog or haze formation. Atoms and molecules of atmospheric gases form very small aggregates known as clusters or nanoparticles. Several decades ago, the principle of gas–phase aggregation became a basis for a new technology to synthesise atomic and molecular clusters in laboratory conditions for specific research applications [1,2]. Since then, this technology has been gradually developing into a widely used approach getting a significant impetus in the 1990s and thereafter due to a high relevance for the rapidly progressing nanoscience and nanotechnology field [3–6]. Different types of gas aggregation sources, which are currently commercially available, provide a number of advantages compared to other physical and chemical means of nanoscale synthesis, allow tailoring nanoparticle parameters and assembling them into functional systems, which are of very high demand in various research and industrial branches [7,8]. In recent years, a lot of activities have been carried out on improving the performance and capabilities of the gas aggregation sources as well as related systems for the manipulation of cluster beams [9,10]. Many studies look into the physics of cluster aggregation and the key parameters affecting their formation, thus, paving a way for control of their composition, shape, size and structure [11,12]. A huge amount of research is devoted to applications of gas-phase synthesised nanoparticles as building blocks of functional nanomaterials and devices for optics, catalysis, sensing and imaging, biotechnologies and other fields [13].

Our intension with this special issue was to address recent advances in gas-phase aggregation techniques, trends in synthesis and functionalization of nanoparticles, as well as applications of cluster beams in preparation of functional nanomaterials or modification of surfaces on the nanoscale. Overall, the book provides the reader with a variety of topics within the field: from the technology on the formation of core@shell nanoparticles to the applications of nanoparticle-assembled matrixes and surface modification on the nanoscale. This diversity shows a many-sided interest in the field of nanoparticle gas aggregation and cluster beams.

The collection is started with a review by Popok and Kylián [14], which analyses the state of the art regarding the synthesis of nanomaterials using gas-phase aggregation methods and outlines the main application fields such as catalysis, formation of magnetic media, utilization of nanoparticles for sensing and detection, as well as production of functional coatings and nanocomposites. The paper gives a good overview of different cluster-matter interaction regimes and advantages of the cluster beam method from the application point of view. It also addresses a paradoxical situation between the enormous development of the cluster technology branch and the sparse use of cluster sources at the industrial level.

The second paper by Skotadis et al. [15] is also a review on nanoparticle synthesis in the gas phase but with the specific focus on applications in sensing technologies. The article overviews the operation principle of sensor matrixes based on the change of conductivity and then systematically analyses numerous examples of nanoparticle-assembled films utilised for strain-, gas- and bio-sensing.

Magnetron sputtering-based gas aggregation sources are known as unique tools for controlling the composition, structure and shape of nanoparticles. In the article by Soler-Morala et al. [16], the studies of clusters with Co core and Cr shell spontaneously formed under sputtering of CoCr target are reported. In such structures, the interaction between ferromagnetic and antiferromagnetic materials causes an enhancement of the magnetic anisotropy and improves the thermal stability of the small particles, in particular. It is shown that one can reduce the critical size for room temperature superparamagnetism to below 7 nm.

The paper by Milana et al. [17] reports on a new fabrication process of ionogel composites containing gold nanoparticles embedded using supersonic cluster beams. Such nanocomposites show large electrochemical capacitance, considerable conductance and good electromechanical response. The proposed combination of geometry, materials and technology is promising for the fabrication of a new generation of self-sensing microactuators for bio-inspired applications.

The work by Prysiazhnyi et al. [18] gives one more example of application-oriented studies utilizing gas aggregated silver nanoparticles for matrix-assisted laser desorption/ionization (MALDI) mass spectrometry, which is one of the most powerful tools for composition analysis in chemistry and biology. This investigation clearly shows a strong dependence of the signal enhancement on the nanoparticle size. The obtained results, thus, allow to elaborate on the key factors affecting the detection efficiency and develop practical recommendations on the inorganic nano-matrix parameters.

Energetic cluster beams are less studied compared to nanoparticle soft-landing for practical applications. From this point of view, the work by Nikolaev and Korobeishchikov [19] on surface nanostructuring of optical materials by cluster bombardment is of special interest. The obtained results give a considerable contribution to the development of fundamental aspects of cluster-matter interaction depending on the particle energy and angle of incidence.

Finally, we would like to thank the authors for their contributions to this book despite the hard times due to a partial lockdown of research activities caused by COVID-19. We look forward to future exciting developments in this field.

Author Contributions: Conceptualization: V.N.P.; writing—original draft preparation: V.N.P.; writing—review and editing: V.N.P. and O.K. All authors have read and agreed to the published version of the manuscript.

Conflicts of Interest: The authors declare no conflict of interest.

References

1. Becker, E.W.; Bier, K.; Henkes, W. Strahlen aus kondensierten Atomen und Molekeln im Hochvakuum. *Eur. Phys. J. A* **1956**, *146*, 333–338. [CrossRef]
2. Hagena, O.F. Cluster formation in expanding supersonic jets: Effect of pressure, temperature, nozzle size, and test gas. *J. Chem. Phys.* **1972**, *56*, 1793. [CrossRef]
3. De Heer, W.A. The physics of simple metal clusters: Experimental aspects and simple models. *Rev. Mod. Phys.* **1993**, *65*, 611–676. [CrossRef]
4. Haberland, H. (Ed.) Experimental methods. In *Clusters of Atoms and Molecules*; Springer: Berlin, Germany, 1994; pp. 207–252.
5. Milani, P.; Iannotta, S. *Cluster Beam Synthesis of Nanostructured Materials*; Springer: Berlin, Germany, 1999.
6. Yamada, I. Historical milestones and future prospects of cluster ion beam technology. *Appl. Surf. Sci.* **2014**, *310*, 77–88. [CrossRef]
7. Popok, V.N.; Campbell, E.E.B. Beams of atomic clusters: Effects on impact with solids. *Rev. Adv. Mater. Sci.* **2006**, *11*, 19–45.
8. Kylián, O.; Popok, V.N. Applications of polymer films with gas-phase aggregated nanoparticles. *Front. Nanosci.* **2020**, *15*, 119–162. [CrossRef]
9. Huttel, Y. (Ed.) *Gas-Phase Synthesis of Nanoparticles*; Wiley-VCH: Weinheim, Germany, 2017.
10. Cai, R.; Cao, L.; Griffin, R.; Chansai, S.; Hardacre, C.; Palmer, R.E. Scale-up of cluster beam deposition to the gram scale with the matrix assembly cluster source for heterogeneous catalysis (propylene combustion). *AIP Adv.* **2020**, *10*, 025314. [CrossRef]
11. Melinon, P. Principles of Gas Phase Aggregation. In *Gas-Phase Synthesis of Nanoparticles*; Huttel, Y., Ed.; Wiley-VCH: Weinheim, Germany, 2017; pp. 23–38.

12. Polonskyi, O.; Ahadi, A.M.; Peter, T.; Fujioka, K.; Abraham, J.W.; Vasiliauskaite, E.; Hinz, A.; Strunskus, T.; Wolf, S.; Bonitz, M.; et al. Plasma based formation and deposition of metal and metal oxide nanoparticles using a gas aggregation source. *Eur. Phys. J. D* **2018**, *72*, 93. [CrossRef]
13. Milani, P.; Sowwan, M. (Eds.) *Cluster Beam Deposition of Functional Nanomaterials and Devices*; Elsevier: Amsterdam, The Netherlands, 2020. [CrossRef]
14. Popok, V.N.; Kylián, O. Gas-Phase Synthesis of Functional Nanomaterials. *Appl. Nano* **2020**, *1*, 25–58. [CrossRef]
15. Skotadis, E.; Aslanidis, E.; Kainourgiaki, M.; Tsoukalas, D. Nanoparticles Synthesised in the Gas-Phase and Their Applications in Sensors: A Review. *Appl. Nano* **2020**, *1*, 70–86. [CrossRef]
16. Soler-Morala, J.; Jefremovas, E.M.; Martínez, L.; Mayoral, Á.; Sánchez, E.H.; De Toro, J.A.; Navarro, E.; Huttel, Y. Spontaneous Formation of Core@shell Co@Cr Nanoparticles by Gas Phase Synthesis. *Appl. Nano* **2020**, *1*, 87–101. [CrossRef]
17. Milana, E.; Santaniello, T.; Azzini, P.; Migliorini, L.; Milani, P. Fabrication of High-Aspect-Ratio Cylindrical Micro-Structures Based on Electroactive Ionogel/Gold Nanocomposite. *Appl. Nano* **2020**, *1*, 59–69. [CrossRef]
18. Prysiazhnyi, V.; Dycka, F.; Kratochvil, J.; Stranak, V.; Popok, V.N. Effect of Ag Nanoparticle Size on Ion Formation in Nanoparticle Assisted LDI MS. *Appl. Nano* **2020**, *1*, 3–13. [CrossRef]
19. Nikolaev, I.V.; Korobeishchikov, N.G. Influence of the Parameters of Cluster Ions on the Formation of Nanostructures on the KTP Surface. *Appl. Nano* **2020**, *2*, 25–30. [CrossRef]

Review

Gas-Phase Synthesis of Functional Nanomaterials

Vladimir N. Popok [1],* and Ondřej Kylián [2]

[1] Department of Materials and Production, Aalborg University, 9220 Aalborg, Denmark
[2] Department of Macromolecular Physics, Faculty of Mathematics and Physics, Charles University, 180 00 Prague, Czech Republic; Ondrej.Kylian@mff.cuni.cz
* Correspondence: vp@mp.aau.dk; Tel.: +45-9940-9229

Received: 31 August 2020; Accepted: 29 September 2020; Published: 2 October 2020

Abstract: Nanoparticles (NPs) of different types, especially those of metals and metal oxides, are widely used in research and industry for a variety of applications utilising their unique physical and chemical properties. In this article, the focus is put on the fabrication of nanomaterials by means of gas-phase aggregation, also known as the cluster beam technique. A short overview of the history of cluster sources development emphasising the main milestones is presented followed by the description of different regimes of cluster-surface interaction, namely, soft-landing, pinning, sputtering and implantation. The key phenomena and effects for every regime are discussed. The review is continued by the sections describing applications of nanomaterials produced by gas aggregation. These parts critically analyse the pros and cons of the cluster beam approach for catalysis, formation of ferromagnetic and superparamagnetic NPs, applications in sensor and detection technologies as well as the synthesis of coatings and composite films containing NPs in research and industrial applications covering a number of different areas, such as electronics, tribology, biology and medicine. At the end, the current state of the knowledge on the synthesis of nanomaterials using gas aggregation is summarised and the strategies towards industrial applications are outlined.

Keywords: gas-phase synthesis of nanoparticles; cluster sources; nanomaterials prepared by gas aggregation

1. Introduction

The fast-growing field of nanotechnology utilises nanoparticles (NPs) as building blocks for ultra-small systems and nanostructured materials with required, quite often unique, functionality [1–5]. Within a wide spectrum of chemical and physical methods used for NP synthesis, the gas-phase aggregation technologies have reached maturity at the research level, which makes them promising to be transferred to the industrial scale in the coming years [6,7].

Gas aggregation of atoms and molecules is a well-known naturally-occurring phenomenon. The examples are clouds or smoking fires. The condensation requires the lowering of temperature and precursors/germs. These conditions can be created artificially in gas aggregation sources enabling the formation of aggregates of atoms or molecules, so-called clusters [8]. When reaching a size of over several tens of constituents, the cluster dimensions reach the nanometre scale and they are often called nanoparticles. Therefore, in this review words "cluster" and "nanoparticle" are used as synonyms.

We can look back in time and find some single instances for the use of gas aggregation in thin film formation even 90 years ago [9]. In the 1950s and 1960s, the first experiments with small gas clusters were carried out and the concept of electrospray source was introduced [10–13]. In the 1970s, an idea similar to electrospraying was applied for the formation of cluster ion beams of metals, mostly of those with low melting point (Li, Cs, Sn, Ga and Hg); so-called liquid metal sources were developed [14–16]. The first cluster sources based on the adiabatic cooling of expanded gas in supersonic jets [17] and on the gas aggregation of vaporized solids (metals and semiconductors) were also constructed in the 1970s [18,19]. A range of new gas aggregation techniques for cluster formation was developed throughout the 1980s. For instance, laser ablation allowed the extension of cluster production over materials with a high melting point [20,21]. This method led to the discovery of fullerenes [22]. The ion sputtering method was adopted for the production of small clusters [23]. The idea to use arc discharge for cluster formation was introduced [24], which later led to the development of the so-called microplasma source [25]. An important step in the cluster source history was the development of the magnetron sputter discharge technique, which has been known for a long time and used in thin film formation [26], towards NP production at the beginning of the 1990s [27]. Currently, magnetron sputtering cluster sources are among the most popular ones. They are available from a number of commercial suppliers in different configurations and allow for the formation of NPs of different species, structures and geometries [6,7,28,29]. From the 1990s to the 2010s, the gas aggregation cluster techniques have undergone substantial progress due to the high relevance of the rapidly developing nanoscience research [4–7,30–38]. However, to be used in the industry, sources with high deposition rate enabling mass production of NPs are required. One of the recent inventions in this direction is a source based on the assembly of atoms in a metal-loaded cryogenically-cooled rare gas matrix initiated by ion beam impact [39]. A production of ~10 mg of clusters (Au_{100}) per hour has been demonstrated [40].

From the application point of view, the formation of NPs by gas-phase aggregation provides several advantages. Since the clusters are produced in a vacuum from ultra-pure targets, this method allows very good control of composition. The gas-phase synthesis is not limited to homoatomic clusters but allows production of compound NPs, for example metal-oxide ones [28]. Such NPs may be fabricated either by adding a small amount of oxygen into the inert working gas [41,42] or by in-flight oxidation of metal clusters using an auxiliary oxygen-containing plasma before they reach the substrate [43]. Furthermore, recent studies showed that the gas aggregation sources may also be employed for the synthesis of alloy and compound NPs with tuneable shape and structure (spheres, cubes, Janus- and dumbbell NPs, as well as core@shell, multi-core@shell or core@satellite ones) [7,28,44–55]. Adding a mass-filtering system after the gas aggregation stage gives the possibility of carrying out size selection [35,56]. Varying the kinetic energy allows adjusting the cluster-surface interaction mechanism [36,57]. In the cluster sources, NPs are typically collimated into beams. Then, by scanning and varying the exposure time one can control the surface coverage or volume filling factor of clusters [32,58,59] as well as form areas with a gradient in particle density or carry out patterning through masks [35,60–62].

2. Cluster-Surface Interaction Regimes

Cluster-surface interaction can be considered of either low- or high-energy. In the low-energy regime, which is often called soft-landing, the kinetic energy per atom E_{at} is below the binding (cohesive) energy of the cluster constituents. The typical limit of E_{at} for this regime is below one eV. Under such impact conditions, the composition of deposited NPs is preserved but their structure and shape can undergo distortions and deformations.

With the kinetic energy increase, one can approach or slightly overcome the binding energy value. In this case, the energy locally transferred to the substrate can be enough to produce small radiation damage and the NP can become trapped. Such energy regime is called pinning.

Further increase of the kinetic energy leads to an energetic impact: the cluster fragments and the constituents become back-scattered or implanted.

Understanding physics and effects occurring under the above-mentioned regimes plays a crucial role in the rational design of nanoparticle-based materials with appropriate functionality. The main issues for every regime are briefly overviewed below while for a more comprehensive description the interested reader should refer to [57].

2.1. Soft-Landing

The simplest scenario of NP soft-landing may be described by the ballistic-aggregation model, in which the addition of particles to a growing structure occurs via randomly selected linear trajectories [63]. This model, originally developed for the explanation of the structure of colloidal aggregates [64,65], assumes that the arriving particle stays at the place of impact and does not move, at least at the time scale of the new particle arrival at a neighbouring site. This results in the formation of a randomly arranged array of NPs on a substrate material that for the higher fluence transforms into a stacked nanoparticle film shown in Figure 1a. Such mesoporous aggregates have densities as low as about one-half of the corresponding bulk materials densities with functional properties given by the properties of original free clusters that are preserved upon soft-landing. Based on the numerical models, the growth of nanoparticle arrays in this deposition mode is characterised by the universal scaling law:

$$R \sim t^{\beta}, \tag{1}$$

where t is the thickness of the deposit, R is its roughness, and β stands for the growth factor that reaches value of 0.33 in the case of ballistic deposition [66,67]. Naturally, β has to be understood as the value that describes the idealised ballistic-aggregation growth of layers that does not account for the size distribution of incoming NPs, their sticking behaviour or scattering [68]. Despite slight deviations, experimentally determined values of β were found to be independent of the composition of deposited NPs and match well the theoretical value [5,69]. This is of great significance for the practical use as it paves the way for the tailor-made production of nanostructured coatings whose surface roughness may be precisely tuned by the NP fluence.

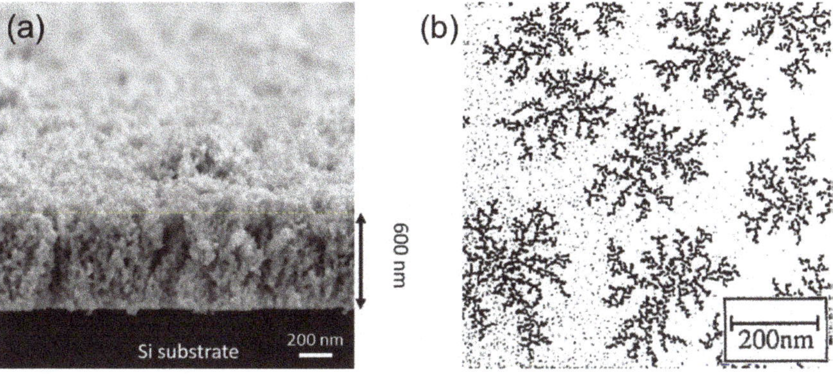

Figure 1. (**a**) Scanning electron microscopy (SEM) image of the cross-section of 600 nm thick film of soft-landed Au NPs. (**b**) Transmission electron microscopy (TEM) micrograph of a 0.5 nm thick antimony film prepared by Sb_{2300} (diameter about 5 nm) cluster deposition on graphite at room temperature. Panel (**b**) is reprinted from [70] with permission from Elsevier.

Obviously, the applicability of the ballistic-aggregation growth model is limited to the cases when the NPs may be considered as immobile after they reach the substrate or growing film. This is, in general, true for particles of a larger mass, transition metal clusters like Co or Fe [34] or in the later stages of the nanoparticle film growth when the new arriving clusters fall on already deposited ones. These become irreversibly immobilized either due to sticking between the NPs or due to the existence of a barrier that prevents them from falling down onto the substrate. However, as proved in numerous studies, even relatively large clusters composed of thousands of atoms may diffuse relatively quickly on a substrate at speeds that are in some cases comparable with those of the surface diffusion of single atoms [70–72]. Based on the experimental and theoretical studies it is believed that the cluster moves on a surface due to a collective mechanism and, thus, diffuses as a rigid structure in a similar manner as a single atom. According to the molecular dynamics (MD) simulations, cluster movement is a combination of its rotation and translation induced by the mismatch of lattice parameters of a substrate and cluster [72,73]. Although the exact mechanism of the diffusion of NPs is still under debate, it strongly depends on the strength of substrate-cluster interaction, temperature or presence of surface defects or steps. In general, the high mobility of NPs was observed on weakly interacting substrates such as highly-oriented pyrolytic graphite (HOPG). The ability of NPs to move on a substrate results in a formation of ramified aggregates (for an example, see Figure 1b), which may be explained by the deposition–diffusion–aggregation (DDA) model. It assumes that a diffusing cluster may either encounter another diffusing NP that leads to the emergence of a new immobile "island" (nucleation event) or might be captured by already existing "island" (growth event) [74]. The actual number and appearance of growing nanostructures are then given by the combination of the particle flux towards the substrate (i.e., deposition rate) and the diffusion coefficient of individual particles.

In the aforementioned description of the DDA model so far only a juxtaposition of colliding particles was taken into the account, i.e., the individual NPs that form an aggregated nanostructure preserve their size and structure. This scenario is, however, in many cases not realistic as the colliding NPs may fully or partially coalesce. This process may be, according to a recent review [75], divided into three consecutive phases: (i) initial contact and formation of an interface between individual NPs; (ii) restructuralization/deformation of NPs supported by the heat release connected with the lowering of the free surface area of interacting NPs [76,77]; and (iii) sphericization induced by thermal diffusion of surface atoms. Naturally, the coalescence process and its rate are strongly dependent on the properties of interacting particles (e.g., their composition, size ratio, shape, crystallinity, the relative orientation between them, degree of order/disorder), temperature as well as on the substrate properties [78]. Depending on these parameters, different final structures of sintered NPs may be observed that range from structures with small neck formed between two clusters, dumbbell-like and ovoid structures to spheres [75].

The possibility of diffusing clusters to coalesce upon mutual contact naturally affects the shape of formed aggregates. A well-known example of this is the variation of the structure of created nanoislands observed in [79] for antimony clusters deposited onto graphite surface that changes from the fractal-like structure to compact islands of spherical shapes as the size of primeval Sb decreases from 500 atoms down to four atoms (Figure 2). Moreover, as the mobility of diffusing NPs is substantially influenced by the surface defects [80–83], the morphology and position of formed nanostructures may be at some extend controlled by the pre-treatment or pre-patterning of the substrate. For instance, it was shown that bends formed on a graphite surface may act as guides for aggregation of Ag_n clusters [84]: while concave bends were found to force clusters to linearly aggregate along the bend, the convex bends behave as repulsive barriers for the particle diffusion that makes it possible to laterally constrain the aggregation in variable extent including the case of quasi-one-dimensional assembly of clusters (see Figure 3a). Another example represents HOPG functionalised by a focused ion beam (FIB) prior to the deposition of clusters. As shown in [85–88], this opens the possibility to fabricate ordered arrays of aggregated NPs (see Figure 3b). Recently, such ordered silver NP patterns have been shown for the FIB-treated graphene [89]. The possibility of Cu_n clusters ordering was reported for the nanorippled Si templates using glancing angle deposition [90]. Finally, it was also suggested to use soft-landed clusters to visualise step edges, grains boundaries, grains orientation or elastic strain fields on HOPG surfaces that are all invisible by SEM [91].

In addition to previously described growth modes, the self-aggregation of clusters into ordered arrays with no mutual contact of individual clusters was reported for gold [92], platinum [92–94], and CoPt [95] clusters. Absence or presence of contact between clusters was found to be dependent on the base pressure during the deposition, size, and reactivity of NPs. For small clusters and deposition pressures above a critical value, surface reactions (e.g., CO adsorption on Pt_n clusters) modify the surface of clusters and such formed interfacial layer represents an effective repulsive barrier that obstructs the contact between individual particles (Figure 4) [92]. Additionally, this effect may even be promoted by the deformation of the substrate surface induced by small Pt_n clusters that may explain the occurrence of arrays of non-contacting Pt_n clusters observed under ultra-high vacuum (UHV) deposition [94].

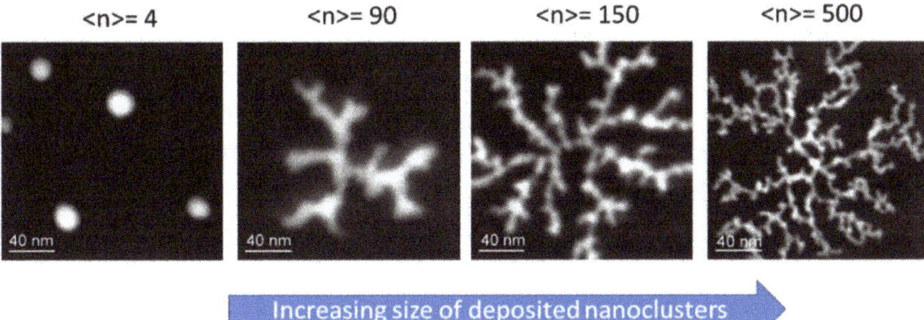

Figure 2. TEM images of antimony islands grown on graphite surfaces for different sizes of the incident clusters. Reprinted from [79] with permission from Elsevier.

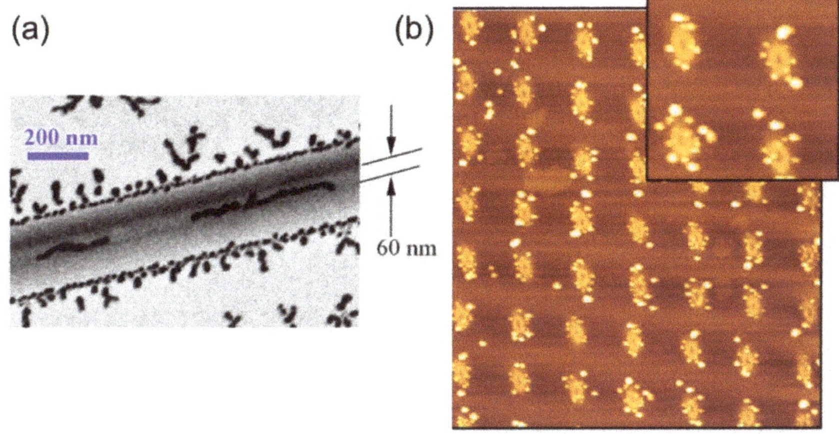

Figure 3. (**a**) Graphite pleats with three faces and two monolayers of silver cluster deposition. The fractal growth on the top face is restricted by the repulsive, convex bends between the faces of the pleat. The silver islands are forced to grow linearly and unbranched (reprinted from [84] with permission from APS). (**b**) Atomic force microscopy (AFM) images showing the morphologies of 0.05 nm thick gold-cluster films deposited on FIB-patterned HOPG substrates at 400 K (reprinted from open-access source [85]).

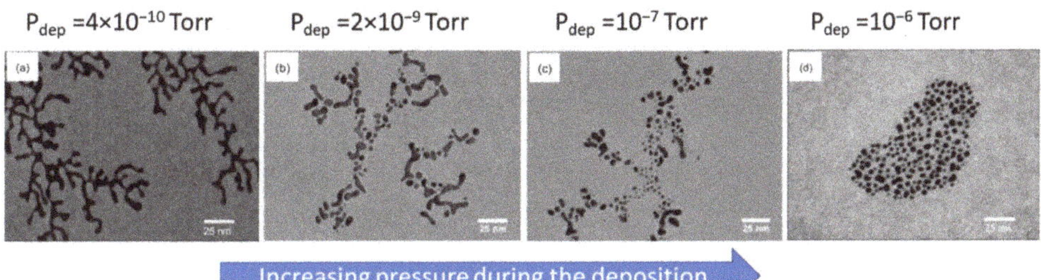

Figure 4. Effect of the pressure (P_{dep}) during the deposition of the size-selected gold clusters (2 nm in diameter) on island morphologies. Reprinted from [92] with permission from APS.

The small NPs can, in addition, undergo important structural (rearrangement of cluster constituents) and morphological (mainly flattening) changes when in contact with a substrate [96–98]. This is due to the dominant role of cluster-surface interaction with decreasing the size of deposited particles. The degree of such changes is dependent on the surface and interface energies as well as on the binding energies of cluster constituents. For instance, according to MD simulation of Au_{440} cluster deposited onto an Au(111) substrate, the complete epitaxy is achieved within 100 ps, as shown in Figure 5. At this point, it has to be stressed that the largest cluster size leading to full contact epitaxy upon deposition decreases rapidly with the decreasing temperature. To give an example, Cu NPs deposited onto Cu substrate at room temperature reached full epitaxy only when they are composed of less than approximately 200 atoms [99]. Due to this, larger NPs remain in a non-epitaxial configuration at room temperature, although the atomic planes in the vicinity of a substrate may become epitaxial with it.

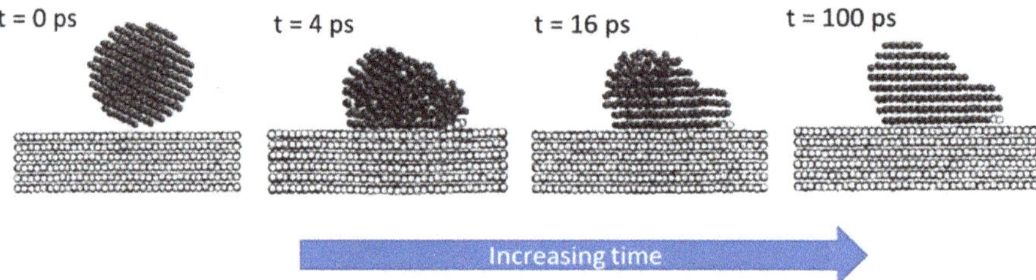

Figure 5. Snapshots of the deposition of an Au_{440} cluster on an Au(111) surface. Reprinted from [96] with permission from APS.

For completeness, it is worth mentioning that clusters may even be rebounded from the surface at relatively low impact kinetic energy. This effect, which was reported for clusters that undergo plastic deformation on impact [100], is highly sensitive on the strength of cluster-surface attraction C. As shown by MD simulations, the transition between adhesion and reflection of nanocluster occurs as the ratio of the kinetic energy to the adhesion energy (so-called Weber number) passes through unity [101]. From the practical point of view, changing the actual value of C (e.g., by exposure of a substrate to electron beam) enables to control the adhesion/reflection of nanoclusters on different positions on a substrate even at the nanometer level. This, in turn, allows for the production of well-defined patterns of clusters needed for the fabrication of functional nanodevices [100,102]. An example of this is depicted in Figure 6, where the SEM image of Bi cluster-assembled wire with an average width of ~50 nm, supported on a SiO_2 passivated Si substrate, is presented.

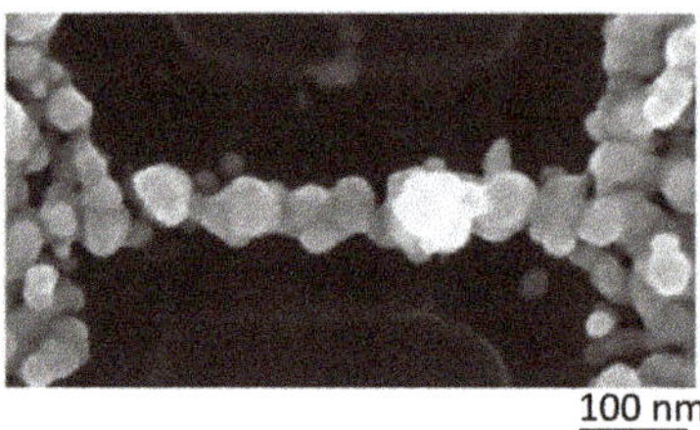

Figure 6. SEM image of Bi cluster-assembled wire supported on a SiO_2 passivated Si substrate. Reprinted from [100] with permission from AIP.

2.2. Pinning

The increase of the kinetic energy on interaction with a surface can cause deformation of the cluster shape and even a partial fragmentation if this energy approaches the binding energy of the cluster constituents. At the same time, the energy transfer to the substrate can lead to point defect formation and the cluster can become trapped [103]. This type of interaction is called pinning because it causes surface immobilization of the cluster. Pinning of various metal clusters (Ag_n, Au_n, Pd_n, Ni_n and Co_n) is mostly studied on HOPG due to atomically flat surface [103–108]. Modelling of the metal cluster interaction showed that the energy required for the carbon recoil production E_r is approximately 5–6 eV [108,109]. Thus, above this threshold value, one can expect the cluster pinning to occur. But it was also found that the pinning energy E_{pin} depends on cluster species, size and substrate material. The following empirical equation was suggested [109]:

$$E_{pin} = \frac{nmE_r}{4NM},$$ (2)

where n is the number of atoms in the cluster, m is the atomic mass of the cluster constituents, N is the number of recoil atoms and M is the atomic mass of the target.

The minimal energies required for pinning have the same order of magnitude with the cohesive energies of metal clusters. Hence, one can expect that a cluster can remain intact or may just partly fragment under the pinning regime. However, as found by MD simulations and confirmed by scanning tunnelling microscopy (STM), the resulting cluster shape significantly depends on species; Ni_n clusters are more compact, while Ag_n and Au_n clusters have tendencies to spread out forming flat structures on pinning [110]. In Figure 7, one can see that Co_n clusters follow the same tendency as Ag_n and Au_n, forming a flat island with a few atoms penetrating beneath the top graphene layer. The STM experiments revealed the formation of one atomic-layer-thick small cobalt islands at the places of impact [107], thus, supporting the modelling.

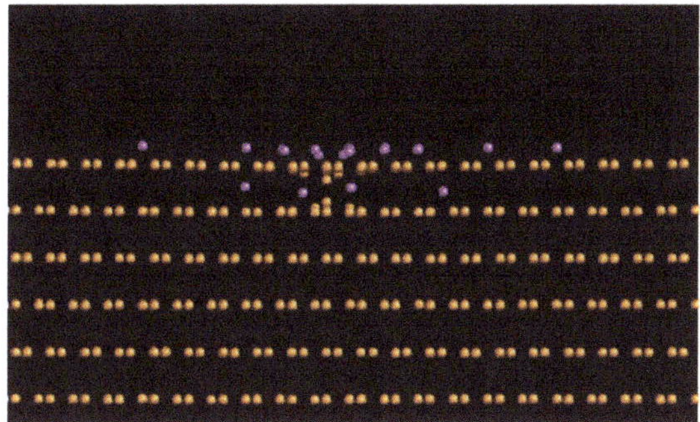

Figure 7. MD simulation, viewed in cross-section, of Co_{30} cluster 2 ps after the impact with the energy of 300 eV/cluster (10 eV/atom) on graphite.

One of the interesting consequences of metal cluster pinning to graphite is the possibility of nanoscale etching. It was found that the annealing at temperatures 600–650 °C in ambient atmosphere leads to the formation of the pits at the cluster location [108,111]. These pits typically have a hexagonal or circular shape (see Figure 8a) that is assigned to the local oxidative etching of the damaged on impact areas. Depth of the pits correlates well with the depth of radiation damage caused by the cluster impact [108,111,112]. To attack the underlying pristine graphene planes the etching time must be significantly increased. For the pinned Co_n clusters it was also found that the heating can cause worm-like planar trenches as can be seen in Figure 8b [107]. High temperature increases diffusive mobility of the residual clusters and, at the same time, they can serve as catalyst particles promoting reactions of atmospheric oxygen with carbon as schematically illustrated in Figure 9. The heating induced channelling of graphite and graphene by NPs deposited by other than gas aggregation techniques has been also reported and even suggested as a "catalytic pen" method for high-precision lithography [113–115]. However, it does not seem that there is an easy way to control the direction of particle motion and, therefore, this approach has not received any further development.

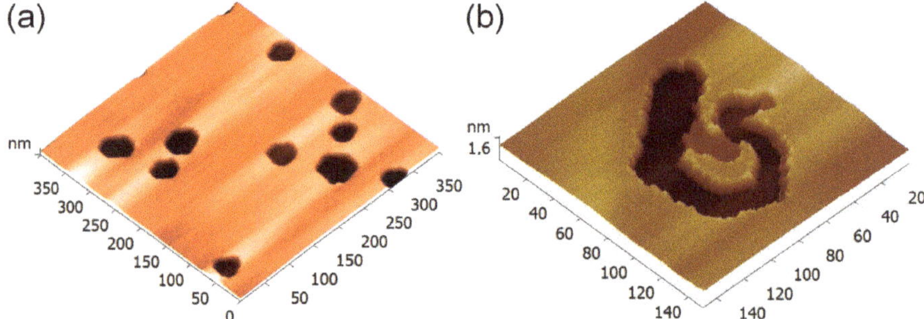

Figure 8. STM images of graphite after the impact of (**a**) Co$_{63}$ clusters with the energy of 130 eV/atom (depth of pits ca. 3.5 nm) and (**b**) Co$_{50}$ cluster with the energy of 12 eV/atom (channel depth is 1 (lighter) and 2 (darker) graphene layers). In both cases, the samples underwent thermal-induced etching (600 °C, 3 min in the ambient atmosphere).

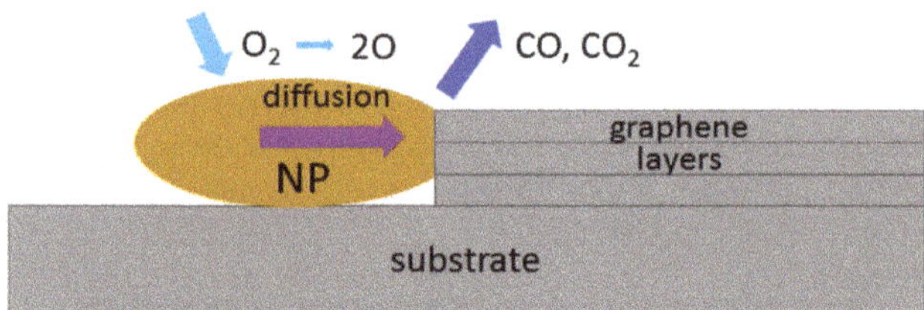

Figure 9. Schematic picture of a NP making a trench in graphite/graphene by diffusing and catalysing the reaction of atmospheric oxygen with carbon atoms.

2.3. Sputtering and Implantation

The energetic collision of an ion with atoms of a target can cause a knock-on effect leading to either sputtering of the atom or forming a recoil, which moves inside the target producing a displacement cascade [116,117]; even a small crater can form [118,119]. Crater formation was found to be a much more pronounced phenomenon for a cluster impact where the multiple collision (almost simultaneous interaction of many cluster constituents with a number of target atoms) leads to a high-energy transfer very locally. At the central part of the collision area a depression is formed while the atoms from the periphery of the impact area possess momenta directed away from the surface (see Figure 10a) [120]. These atoms either become sputtered or produce an atomistic flow leading to the rim formation (see Figure 10b). The crater parameters depend on the cluster size and energy as well as on the properties of a target. A variety of different target materials and cluster species have been studied both theoretically and experimentally [121–131]. The generally-found tendency is that the materials with higher density and melting point are less favourable for the crater formation. One of the examples is diamond [132]. Another allotropic form of carbon, graphite, responds to the cluster impacts very elastically. As shown by the MD simulations, the crater can be formed at the initial stage of impact but then the graphene sheet oscillations cause its closure [133].

Along with craters, hillocks (nm-scale protrusions, typically of cone-like shape) are often observed [123,134,135]. The MD simulations of keV-energy cluster impact predict compression of the material up to GPa level leading to temperature rise up to a few thousand degrees [120,136]. This temperature level is above a melting point for any material. The molten matter can be pushed out and quenched forming a hillock. The origin for the expulsion can be the difference in density or tension between the hot melt and the surrounding solid as well as an elastic rebound of the bulk because of the initial compression and shock wave formation. In some cases, a hillock can appear in a centre of a crater (see Figure 11) as, for example, found for the Ar$_n$ cluster impact on silicon and sapphire [128]. By analogy with meteorite impact craters on a planet surface, where a centrally-located bump is often seen [137], these structures are called complex craters. For more details about crater formation on cluster impact, one can refer to [37].

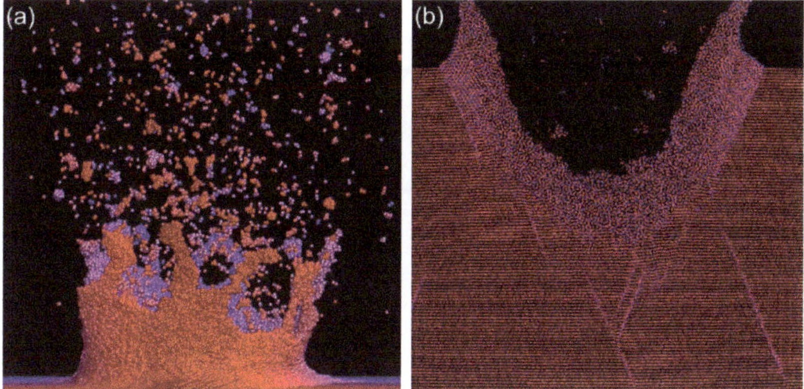

Figure 10. Snapshots of MD simulations of 650 eV Au$_{13}$ cluster collision with a gold target: (**a**) sputtering of atoms at the beginning of the collision and (**b**) crater formation after 16 ps. Reprinted from [37] with permission from Elsevier.

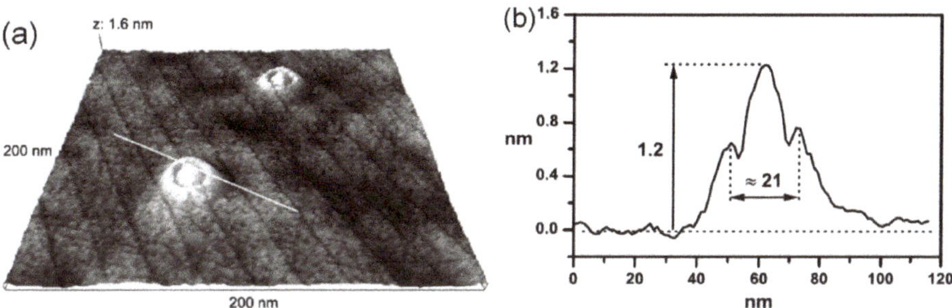

Figure 11. (**a**) AFM image of complex craters formed on the surface of silicon after 18 keV Ar$_{12}^{+}$ cluster impact and (**b**) cross-sectional profile of one of the craters. Reprinted from [128] with permission from Springer.

As mentioned above, many target atoms get momenta directed away from the surface on the cluster impact, i.e., provide a very pronounced lateral effect. Even at normal to the surface incidence, the distribution of the sputtered atoms has a strong angular dependence and the sputtering yield Y does not follow so-called cosine law with respect to the emission angle of the atoms Θ, $Y(\Theta) \sim \cos^{m}\Theta$ with $m \geq 1$, which is typical for monomers [138]. Recent studies of the Cu and Mo sputtering by argon clusters have shown an anisotropy of atom distribution over azimuthal angle, which is found to be related to the crystal orientation and cluster size [139]. The use of clusters increases Y from a few atoms typical for monatomic ions to tens and even hundreds [140] making the cluster beams to be an efficient tool for sputtering and smoothing of surfaces [141]. Additional advantages of cluster sputtering in comparison with ion or plasma assisted processing are short-range damage, high spatial resolution and elimination of charge accumulation on the surface. These properties promoted the utilization of cluster ion beams in the field of secondary ion mass spectroscopy (SIMS), especially in application to organic materials [142,143]. The C$_{60}^{+}$ and Ar$_{n}^{+}$ cluster ion beams have been primarily used providing not only depth profiling of the organic species but also imaging of biological tissues [144–147].

Since clusters generate multiple collisions with target atoms on impact, the physics of cluster implantation has several fundamental differences compared to a monatomic bombardment. At the beginning of impact, the cluster kinetic energy and momentum are locally transferred to the target causing a compression up to GPa level and rapid heating up to 10,000–100,000 K [120,123,136]. The pressure and temperature rise lead to the development of shock waves and significant thermal spikes. The latter can cause local material melting along the cluster path [148]. Under typical implantation conditions, the kinetic energy per cluster constituent is considerably higher than the cluster cohesive energy. The aggregate becomes broken on impact and the penetrating into the target constituents form collision cascades of a complex nature due to the non-linear effects originated by the interaction of these cluster constituents not only with the target atoms but also with each other. One of the closely related phenomena is the so-called clearing-the-way leading to deeper implantation of the cluster atoms compared to the monatomic projectiles with the same incident velocity [149,150]. It is explained by pushing the target atoms by the "front" cluster constituents out of the way of the cluster "rest", thus, reducing the total stopping power. The clusters composed of heavy elements cause more clearing of "light" targets and, contrarily, the effect is negligible if the cluster consists of light atoms but the target is represented by heavy elements [151]. This phenomenon is well illustrated by metal cluster implantation into polymers. It was found that small (ca. 2.7 nm in diameter) Pd NPs impacting poly(methyl methacrylate) (PMMA) with the energy of only 0.5 eV/atom become dispersed to the depth of 50 nm [152]. The clusters also stay intact because the kinetic energy per atom is below the cohesive cluster energy. Such deep penetration cannot be explained exclusively by the clearing-the-way effect. Using MD simulations, it was suggested that the local thermal and pressure spike generated by the collision facilitate cluster embedding due to an increase of chain mobility [153]. This approach on shallow implantation of metal clusters into polymers paves a way for the formation of composite films with tuneable properties towards diverse applications, which will be addressed in Section 3.4.

An important issue for cluster implantation is the prediction of the projected range, R_p. This problem was successfully solved for monatomic ions using the binary collision approach. Nowadays, well-developed and experimentally verified simulation codes exist, for example, SRIM or TRIDYN [154,155]. However, due to the complexity of the cluster-matter interaction, the simulations of cluster implantation show diversity in scaling laws with respect to energy $R_p \sim E^{1/3}$ and $E^{1/2}$ [156,157] as well as to cluster size (number of atoms) $\sim n^{\alpha}$ where α ranges between 0.31 and 0.45 [157,158]. Unfortunately, it is difficult to conclude about the correctness of the suggested laws because the experimental data on systematic measurements of R_p are rather limited, except the cluster implantation into graphite.

The specificity of graphite is in its layered structure with weak Van der Waals bonds between individual graphene sheets but strong covalent bonds in the planes. The implanted cluster creates damage, amorphizing the material along the track to a certain depth. The volumes of amorphous carbon created by individual cluster impacts can be etched by short-term heating in the presence of oxygen forming pits as described in Section 2.2 (see Figure 8a). Using AFM or STM one can find the depth of radiation damage as a function of implantation energy. For not too high energies (keV scale), this depth corresponds well to R_p [132,133]. The experiments with different cluster species implanted into HOPG allowed to combine the scaling laws for energy and cluster size (mass) using the cluster momentum p, which is linearly proportional to the mass, and the velocity be a square root function of energy ($E^{1/2}$). This approach led to a new universal scaling law $R_p \sim p$ proved for a wide range of cluster species and energies (see Figure 12) [108,111,133,159]. A similar dependence was also found for the keV-energy argon cluster ions implanted in diamond [160].

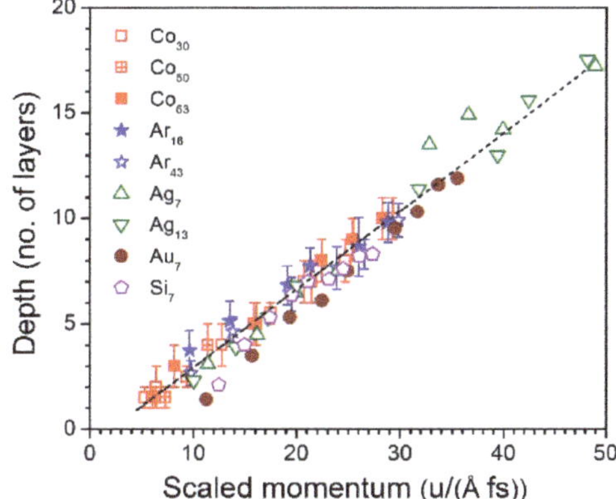

Figure 12. Experimentally-found depth of pits (in number of graphene layers) as a function of scaled momentum (momentum per unite of cross-sectional area of cluster-surface interaction) for different cluster species and sizes. The dashed line represents the best linear fit. The dependence is based on the data published in [108,111,133,159].

3. Functional Nanoparticle-Based Materials

Interest to the practical use of NPs, nanoparticulate films and composites containing NPs comes from different fields. Small particles have a high ratio of surface-to-volume atoms, thus, becoming ideal objects to exploit properties of the outer atoms/surface sites, which play a key role in catalytic applications [161,162]. An ability to control the particle size and composition brings new routes for tuning magnetic properties and creating novel magnetic media [33,163,164]. The phenomenon of localised surface plasmon resonance (LSPR), giving rise to characteristic optical extinction and strong local electromagnetic field enhancement [165], is currently of significant practical interest for applications in sensing, detection and imaging, as well as for optical components and collar cell technologies stimulating the development of novel and sophisticated NP-based systems [166–170]. Controllable assembling of NPs into films or composites allows for conductivity tuning from the variable range hopping to metal-like charge transfer mechanisms opening a way for the fabrication of unique electronic and electro-mechanical devices [171–175]. The NP-filled materials are also of high interest for microwave technologies, bactericidal coatings and membranes, as well as for the formation of surfaces with controlled morphology, wettability and adhesion [176–179].

3.1. Catalytic Applications of Size-Selected Clusters

In sub-nanometre particles, the majority of atoms is located on the surface. These atoms are low-coordinated, thus, promoting a chemical activity. One such example is gold, which is known to be a relatively poor catalyst in bulk but showing considerable catalytic support on the nanoscale in a number of reactions [180,181], for example, oxidation of H_2 and CO, just to mention one of the first publications on this topic [182]. Shrinking the NP size below a hundred atoms also brings into play the quantum phenomena changing the electronic structure and, thus, affecting the chemical properties [183]. Varying the size on the scale of ±1 atom can dramatically change the catalytic efficiency of very small clusters [184,185]. The surrounding medium or supporting substrates are also found to play an important role; metal clusters deposited on oxide surfaces strongly change the catalytic properties due to a charge transfer that, for instance, turns inactive gold and palladium NPs into active catalysts [162,184,186,187]. Tiny Cu_n clusters is another example. In contrast to the gas-phase Cu_n clusters, which are more difficult to oxidize the smaller they are, the supported on amorphous alumina Cu_4 were found to be oxidized easier compared to Cu_{12} and Cu_{20} [188]. This can be seen in the theoretically-obtained phase diagrams presented in Figure 13. The predictions were proven by experimentally-measured temperature at which the transition from $Cu_nO_{n/2}$ to Cu_n occurs in a H_2-rich environment [188]. The studies of small metal clusters and also of metal alloys have shown that expensive metals can be substituted by cheaper ones while keeping the catalytic efficiency at the same level [161,189]. Since the literature on the catalytic studies of supported metal NPs is numerous, we will overview only the advantages of the cluster beam technique for this field.

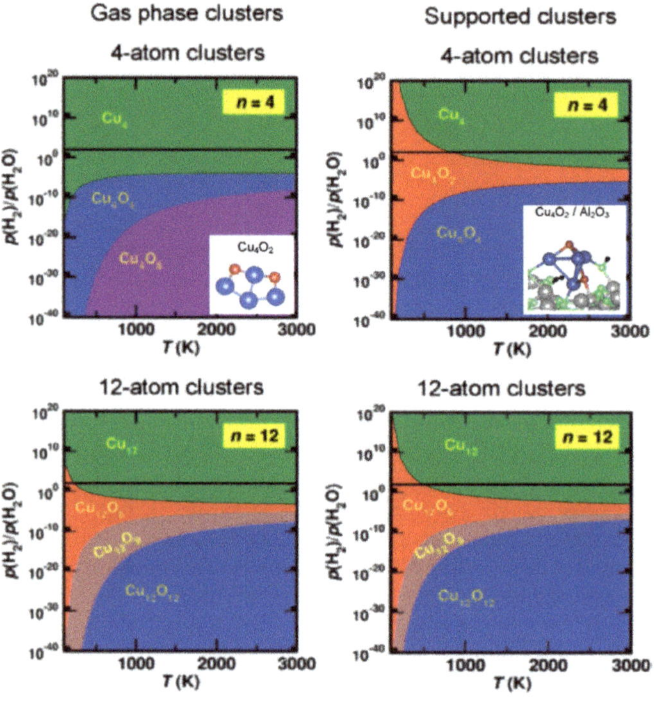

Figure 13. The ratio of the partial pressure of H_2 and H_2O for gas-phase (**left**) and supported on amorphous alumina (**right**) Cu_4 and Cu_{12} clusters as a function of temperature. The green, red and blue areas indicate Cu_n, $Cu_nO_{n/2}$ and Cu_nO_n phases, respectively. Inserts show the calculated lowest energy structures for Cu_4O_2 clusters. Reprinted from [188] with permission from Wiley-VCH.

The main advantages are (i) the size selection to single atom precision for the clusters consisting of up to 2–3 tens of atoms [56,190]; (ii) formation of binary and ternary NPs, as well as core@shell structures [6,44,47,48,191]; and (iii) deposition on any desired substrate with controllable coverage [60]. These pros make the gas aggregation techniques an advantageous tool for the study of model catalysts.

The oxidation activity of tiny Cu_n clusters has been mentioned before. These clusters were also found to be efficient in CO_2 conversion to methanol, $CO_2 + 3H_2 \geq CH_3OH + H_2O$. The turnover rate of about 4×10^{-4} molecules per copper atom and second was reached [192]. Interestingly, the highest rate was found for Cu_4/Al_2O_3 exceeding the activities of other small supported clusters. Strong size-selective reactivity was also observed in CO oxidation on sub-nanometre Ag_n clusters supported on titania [193]. More examples of size-selectivity of various types of model catalyst NPs can be found in [161,189]. It is important to stress that the use of ultra-small NP acting as individual catalytic sites significantly reduces expenses, especially for precious metals. The formation of core@shell structures with cores of less expensive metals and shells composed of active species can also have a considerable economic effect. Generally, core@shell NPs is a class of materials that has recently received increased attention due to a number of advantages for catalysis [194]. In particular, synergetic effects caused by core-shell electron exchange can considerably increase catalytic activity as, for example, found for Pt-Pd systems [195]. Since multiple-target magnetron sources allow precise control of the core and shell growth in one step [6,47], the gas aggregation approach brings large benefits in the development of new model catalysts. Similar to core@shell structures, alloys can also exhibit a synergistic effect enhancing catalytic activity. For example, Pd_6Ru_6 clusters showed far superior performance in alkene hydrogenation compared to the monometallic analogues of Pd and Ru, as well as other particles like Cu_4Ru_{12}, which were also used to catalyse this reaction [196]. Ag_9Pt_2 and Ag_9Pt_3 bimetallic clusters supported on alumina were observed to be more stable while enabling the same catalytic efficiency of CO oxidation as pure Pt_n [197]. Mass-selected NiFe NPs were found to be perfect model systems providing fundamental insights into the water splitting reaction [198]. Engineering of gas aggregated Ni-Mo-S NPs allowed for tuning their morphology, i.e., the abundances of edge sites, thus, affecting catalytic activity, for example in the removal of sulphur from fossil fuels [199].

Disadvantages of the gas-phase aggregation methods for industrial catalytic applications are the requirement to vacuum and relatively low production rate of clusters sources. A typical cluster beam setup enabling the formation and deposition of sub-nanometer particles would produce below 1 microgram of catalyst per hour [4]. However, the recent development of a new approach for NP gas aggregation, so-called matrix assembly cluster source, brings hope for a breakthrough, as it promises to reach the gram scale in the production of supported size-selected metal NPs [40].

3.2. Magnetic NP-Based Media

The origin of magnetic ordering in the solids is based on the exchange interaction of electrons and spin imbalance. Compared to the solid-state, clusters can represent different symmetry (or broken symmetry) and reduced coordination of surface atoms causing a local increase of moments. For the atomic aggregates of Fe, Co and Ni, known as ferromagnetic materials in bulk, a significant increase of spin moment was found with decreasing the size; also a contribution of the orbital moment can become considerable [200,201]. Free clusters of other transition metals, like Rd_n, Ru_n, Pd_n, Gd_n and Cr_n, were also observed to exhibit enhanced magnetic moments [202,203]. However, analysis of magnetic phenomena of free/isolated clusters is out of the scope of this paper and the reader can be addressed to extensive reviews, for example [204,205].

Clusters of 3d ferromagnetic metals were also found possessing enhanced magnetic moments while being supported on a substrate or embedded into a matrix [164,205,206]. Larger moments were observed on weakly-coupled surfaces like insulators, sp-metals or metals with filled d-band, i.e., some type of hybridization between the cluster and host atoms took place. This phenomenon is found to be important for alloy and core@shell clusters representing systems of 3d metals with other transition metals like Pt, Pd, Au, Ag and W [206–208]. Although many found dependences of magnetic properties on cluster size, species and supporting/surrounding matrix still require a more thorough study, the cluster beam method showed excellent capabilities for varying the magnetic properties on the nanoscale. For instance, core@shell particles can be created in the gas aggregation in one step considering the liquid (L) and solid immiscibility of given components A and B (see Figure 14a) and the difference in surface energy (see Figure 14b). Alloying can undergo for small particles, while with the size increase the positive enthalpy of mixing can lead to the phase separation upon condensation (see Figure 14c) and, thus, giving a possibility to control the particle properties including the magnetic ones [208]. Varying the cluster composition/structure or filling factor of NPs in the matrix allows for tuning the coercive force, saturation magnetization and blocking temperature of superparamagnetic/ferromagnetic transition [164,209,210]. The latter is one of the main suppressions for use of small NPs in high-density memory devices; with decreasing size, one becomes confronted by the superparamagnetic limit. A possible solution could be the use of antiferromagnetic surrounding to enhance the anisotropy of a ferromagnetic NP [211]. Apart from the use of clusters in the formation of materials exhibiting giant magnetoresistance, the gas aggregation method provides excellent capabilities to produce granular films with enhanced tunnelling magnetoresistance [212].

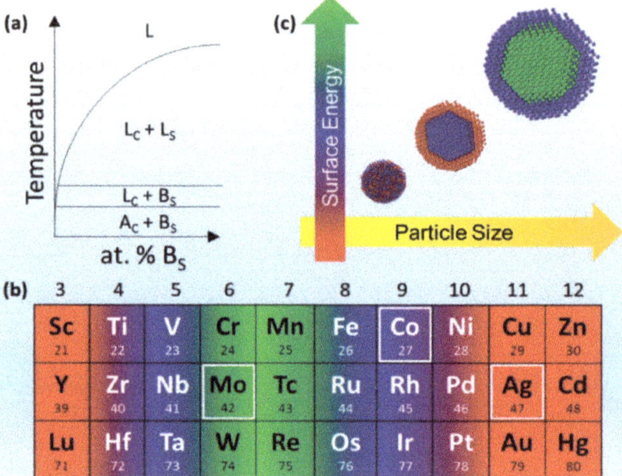

Figure 14. (a) Schematic phase diagram of a system with liquid immiscibility, where A_C and B_S are the core and shell materials, respectively. (b) The relative surface energies for groups 3–12 of the periodic table (green > blue > red). (c) Design concept summary: larger particles with different surface energies will form core-shell structures and blue elements (from groups 4, 5, 8, 9) can be placed in either the core or the shell. Reprinted from [208] with permission from Wiley-VCH.

When forming layers of supported NPs or composites with embedded ones, the surface/volume fraction plays an important role. For an individual cluster of ferromagnetic material, a uniaxial anisotropy with a random orientation of an easy magnetization axis is typical. Decreasing the distance between the NPs to a few lattice constants causes an overlap of the surface electronic wave functions and exchange coupling. Further increase of cluster surface coverage or volume fraction leads to the enhancement of the susceptibility and increase of coercivity [213] until percolation. Beyond the percolation threshold, a correlated super-spin state can be reached. This situation can be described using the parameter characterizing the relative strength of the anisotropy H_r and exchange H_{ex} magnetic fields, $\lambda_r = H_r/H_{ex}$ [163]. For $\lambda_r > 1$, the contribution of the exchange field is small and magnetic vectors of individual particles have random orientation following intra-particle anisotropy axes. With decreasing intra-particle anisotropy, the magnetization of neighbouring particles becomes nearly aligned forming a correlated super-spin but the deviations from perfect alignment can cause a smooth rotation of the magnetization within the magnetic correlation length proportional to $1/\lambda_r^2$. Such superparamagnetic layers show a significant increase of susceptibility and very low coercive filed at room temperature compared to isolated Fe or Co clusters.

NPs with superparamagnetic properties are of wide interest in biological applications: drug delivery, bio-sensing and bio-separation [214]. However, NPs (Fe, Co or allows with other metals) produced by chemical means or physical methods others than gas-phase aggregation are typically efficient enough while being cheaper and simpler. Hence, to our best knowledge, the cluster beam technique is not much in use in this field.

3.3. NPs for Enhanced Sensing and Detection

Cluster assembled nanostructures, which can be obtained using a combination of standard lithography technique and cluster beam deposition, represent a simple method for the formation of electrical contacts and electronic components on the nanoscale. One of such examples is shown in Figure 6, where a nanowire is produced by Bi_n cluster deposition using a mask. It was found that the percolation threshold ensuring the formation of conductive passes through the assembly can be shifted to very low surface coverages of NPs [215]. Controlling the coverage allows tuning the resistance of the assembled metal nanostructures [174]. This capability was used for the fabrication of passive electronic components (resistors and capacitors) by patterned cluster deposition on paper [61]. An example of resistor arrays formed by gold NPs on plain paper is shown in Figure 15. An advantage of the cluster beam technique is in a wide choice of species (both metals and semiconductors) for NP formation.

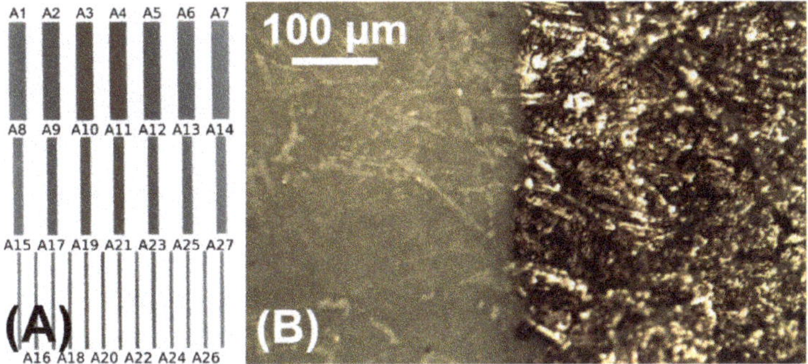

Figure 15. (**A**) Photograph of a set of resistors printed in a one-step process with gold NPs on plain paper. Height of stripes (resistors) is 30 mm, width varies from 1 (bottom) to 5 (top) mm. (**B**) Optical microscopy image of an individual resistor border. Reprinted from [61] with permission from AIP publishing.

Deposited NP assemblies or films were also found to be sensitive to the environmental conditions, i.e., changing the conductance under exposure to some gases. One of the examples are tin oxide nanoclusters (between 3–10 nm in diameter) prepared as a few monolayer thick films which were shown to be highly and fast responsive to hydrogen and ammonia at relatively low temperatures (80–200 °C) by a considerable change of the resistance [216]. Hydrogen sensors based on percolation and tunnelling conductance were also fabricated using Pd_n clusters [217]. In [218], a systematic approach for fabrication of sensor batches on the microscale utilizing hotplates was suggested: NPs of different metal oxides, such as SnO_2, TiO_2, WO_3, Fe_2O_3 etc. could serve for gas sensing.

Cr NPs deposited between silver electrodes fabricated on the polyethylene terephthalate (PET) film were shown to be efficient as strain sensors [175]. A compressive or tensile strain applied to such film changes the inter-particle distance modifying the percolation paths and, thus, the resistance. The sensors have shown much higher gauge factors compared to typical metal foil-based analogues especially for high values of applied strain where the foil-based gauges fail due to cracking. Recently, such approach based on deposited gas aggregated Pd NPs has been suggested for the formation of pressure sensors (see Figure 16) with a resolution as high as 0.5 Pa [219].

Another very widely used method of enhanced sensing is based on the utilisation of LSPR in metal (mostly coinage, Au, Ag and Cu) NPs [165,220]. The cluster beam technique is a very efficient tool allowing the formation of very pure particles with enhanced chemical stability, controlled size and surface coverage, thus, bringing great capabilities for plasmonic property tuning. It has been shown that change of inter-particle distance with increasing surface coverage leads to stronger near-field coupling among metal NPs, thus, allowing for precise control of the plasmon band spectral position. A nearly linear dependence of the band maximum wavelength on the deposition time between ca. 390–570 nm has been demonstrated for the soft-landed gas aggregated Ag NPs [221]. High purity and perfectness of the NP crystalline structure were found to be key factors for strong chemical resistance leading to long-time stability of the plasmonic properties in ambient conditions, which is a very important property for the utilisation of NP-based matrices in surface-enhanced Raman spectroscopy (SERS). Such long-term stability was demonstrated for the gas-aggregated silver [222] and copper [223] NPs (see Figure 17). For the latter, the LSPR stability was additionally promoted by an UV-ozone treatment, which led to the rapid growth of an oxide shell protecting the NP against gradual atmospheric oxidation.

The use of Al NPs is important for some applications where the LSPR resonance is required in UV spectral interval. However, Al is prone to oxidation in ambient air. Therefore, NPs must be embedded in a matrix. Such a strategy was realised in [224], where Al clusters were embedded into 50 nm thick C:H matrix prepared under the simultaneous deposition of the plasma polymer. The composite films demonstrated well-pronounced plasmon bands with maxima between 255–307 nm depending on the particle size. Metal cluster deposition into polymers also opens a way for the production of composite films with the stable tuning of the optical extinction under cyclic compression/stretching [169,225]. One more area of plasmonic properties utilization can be an enhancement of solar cell efficiency. It was recently demonstrated that the deposition of gas-aggregated Ag NPs considerably increases the optical absorption of amorphous Si layers [170].

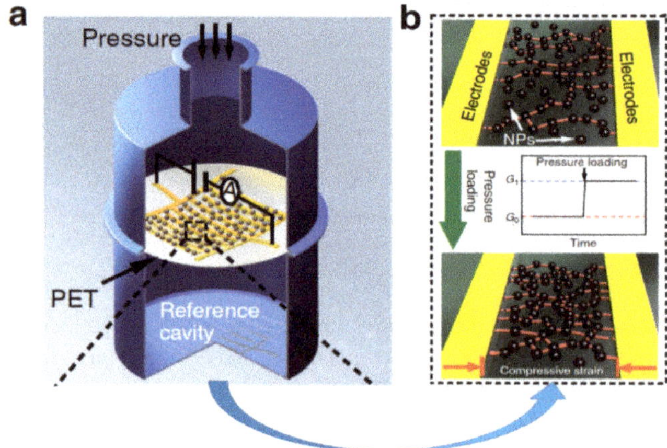

Figure 16. Schematic pictures of a pressure sensor: (**a**) piezoelectric barometric sensor prototype in cross-section and (**b**) operating principle. Red lines linking NPs represent percolation pathways that may exist in arrays. When pressure is loaded, more pathways are generated so that the lines become denser and the resistance decreases. Reprinted from open-access source [219].

Another field in which the use of nanoparticles for enhanced biodetection receives increasing attention is laser desorption/ionization mass spectrometry (LDI MS) [226–229]. In a conventional matrix-assisted LDI (MALDI) technique, the analyte to be studied is embedded into an organic matrix that controls the energy transfer between the incident laser beam and matrix/analyte and allows for the protonation of laser desorbed analyte molecules without their substantial fragmentation [230]. Such produced ionized molecules are subsequently detected by time-of-flight mass spectrometer (TOF-MS). Due to the "soft" ionization process and detection capability of TOF-MS it is possible to detect large molecules with masses up to several hundreds of kDa that makes this technique highly valuable in diverse fields including medical diagnostics, biodefense, clinical chemistry, environmental microbiology, food industry, proteomics or forensic science [231–238]. However, the use of an organic matrix in MALDI possesses critical drawbacks: the poor signal reproducibility due to the inhomogeneous distribution of the analyte in a matrix, limited spatial resolution for imaging or complications in small-molecule detection (the self-ionisation of the organic matrix may interfere with the signal originating from the analyte in the low-mass spectral region). Aforementioned limitations of MALDI may be at least partially overcome if the organic matrix is substituted by inorganic with noble metal NPs (e.g., [239–242]). This method, which is termed as nanoparticle assisted laser desorption/ionization mass spectrometry (Nano-PALDI MS), relies on the ability of NPs to effectively enhance both the desorption and ionization of analyte molecules through the laser-induced heating of the NPs [243] and injection of hot LSPR electrons to the analyte molecule [244]. Although different strategies were developed for the incorporation of NPs into/onto an analyte, the cluster beam deposition technique offers several advantages. Among them, the most important is the possibility to deposit a controlled number of NPs with well-defined properties and high purity onto relatively large areas with good spatial homogeneity. Recently, it has been shown that size of particles matters in Nano-PALDI measurements, thus, emphasising one more advantage of the cluster beam technique [245]. Thus, the features, which are characteristic for the gas aggregation sources, enable to form matrixes for quantitative, reproducible and sensitive detection of small biomolecules [246,247] as well as imaging of lipids in rat kidney [248], heart or brain tissues [249,250].

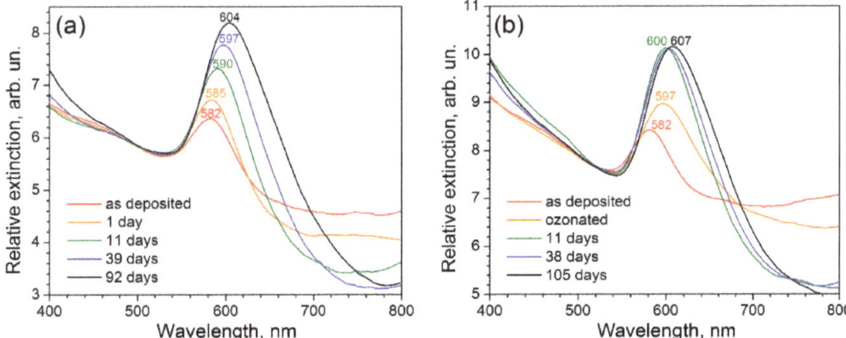

Figure 17. Time evolution of LSPR band for Cu NPs with mean diameter of 19 nm (**a**) kept in ambient air and (**b**) UV-ozone treated for 30 min. and kept in ambient air. Numbers in the panels indicate wavelengths of the band maxima.

3.4. Coatings and Composite Films

Despite the unique properties of NPs and their arrays discussed in the previous sections, one of the limiting factors for their real applications is a weak surface-particle interaction that is a common situation in the case of soft-landing. Due to this, the NP deposits may be easily removed from the surface that hampers their possible use. The high attention is consequently devoted to the development of strategies that enable to fix the NPs on a surface without alteration of their functional properties.

The first approach is based on the implantation of clusters composed of heavy elements into soft materials. As already mentioned in Section 2.3, this may be achieved at kinetic energies of impacting projectiles below the cohesive cluster energy. The typical example represents the embedment of metallic clusters into polymeric substrates by means of supersonic cluster beam implantation (SCBI) developed at the University of Milano [152,251,252]. It has been demonstrated that nanometre size particles may penetrate into a polymeric substrate to the depth of several tens of nanometres while avoiding drastic alteration of the polymer properties typical for NP formation by ion implantation means [253]. Fabricated by SCBI metal/polymer nanocomposites were recently employed as highly deformable elastomeric electrodes with the ability to sustain cyclical stretching [251], mechanical-electro-optical modulators [254], deformable gratings for hyperspectral imaging [255] or stretchable electrodes for the recording of electrical brain activity [252]. Furthermore, such materials are highly promising in the rapidly evolving field of soft robotics [173].

The second method to fix clusters on a substrate utilises their partial or full embedment into a polymeric substrate upon its moderate heating. This approach makes use of the difference in surface energy between metals (in a common situation above 1000 mJ/m^2) [256] and polymers (typically below 100 mJ/m^2) [257]. To minimise the tension at the interface, metal clusters tend to immerse into the polymeric substrate (Figure 18a) until an equilibrium state is reached that is given by [258]:

$$\gamma_{NP} = \gamma_{NP/P} - \gamma_P \cos\varphi \tag{3}$$

where γ_{NP}, $\gamma_{NP/P}$ and γ_P stand for the tension at the interface of NP/air, NP/polymer and polymer/air, while φ is the contact angle as depicted in Figure 18b. In this simplified scenario, that does not account for the forces related to the possible substrate deformation or formation of wetting layer, it is possible to determine the height h of nanoparticle with a radius r above the substrate by [258]:

$$h = r(1 + \cos\varphi). \tag{4}$$

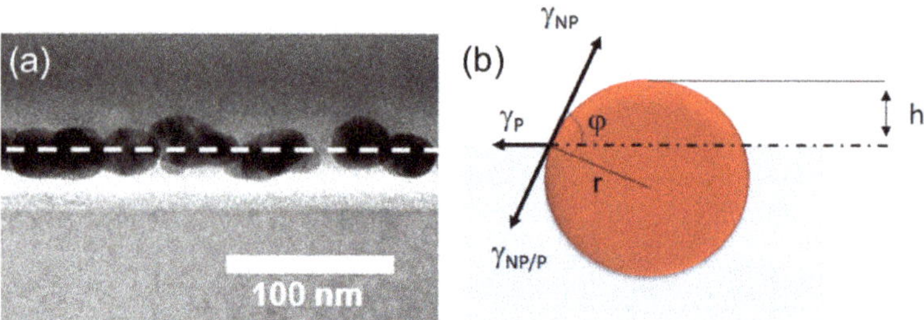

Figure 18. (**a**) High resolution TEM image of Ag NPs partially embedded into PMMA layer with surface indicated by dashed line (reprinted from open-access source [259]) and (**b**) schematic image of NP embedding into soft material (see text for details).

The ability of NP to embed into a soft polymer is substantially promoted when the polymer is heated at a temperature close to its glass transition point at which the flexibility and mobility of polymeric chains increase facilitating, thus, the cluster immersion.

This technique was recently used for the partial embedment of copper, silver and Ti/TiO$_x$ NPs into spin-casted PMMA and polystyrene films [43,258–261]. Such prepared materials may be used either as stable plasmonic-based transducers for protein sensing [259] or as antibacterial coatings [43]. Furthermore, the combination of heat-induced embedding with surface patterning by electron beam lithography represents an interesting option for the facile production of 2D nanoparticle patterns with the desired configuration. In this case, a procedure based on the fabrication of patterned PMMA/polymethylglutarimide (PMGI) coatings, followed by cluster deposition, annealing step and PMMA lift-off (Figure 19) may be employed [261]. An example of such fabricated linear arrays of size-selected Cu NPs is presented in Figure 19 in the dashed circle.

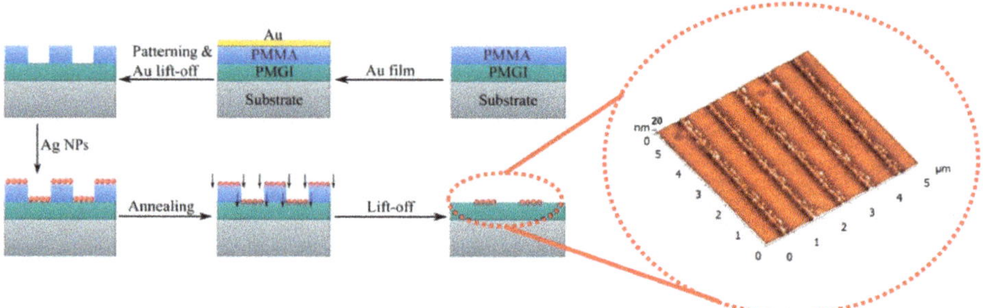

Figure 19. Schematic illustration of the procedure for formation of ordered NP arrays (see text for details) and AFM image of 200 nm wide strips of Ag NPs fabricated by this method. Reprinted from [261] with permission from Cambridge University Press.

NPs may be fixed on a surface also by a thin layer that is deposited over the NP array. Although the overcoat film may be produced by various techniques, the most frequently used are plasma-based deposition techniques (magnetron sputtering, plasma-enhanced chemical vapour deposition, and vacuum evaporation) that are fully compatible with the vacuum nature of cluster beam method and, thus, the deposition of both NP arrays and fixation layer may be performed in a single apparatus. This is a rather important feature as it avoids exposure of NP deposits to ambient atmosphere and, hence, limits possible contamination or modification of reactive particles when in contact with air. Furthermore, the proper selection of the overcoat material may not only provide the desired fixation of NPs but also add a new functional property to the resulting coatings. The representative example of this is the fabrication of antibacterial coatings based on Ag or Cu NPs. In this case, the fixation layer acts also as a barrier that controls the release of bactericidal ions from the coatings into the liquid environment and with it connected antibacterial efficiency of such nanomaterials [262–265].

In addition, the 3D character of NPs provides a roughness to such prepared coatings. The possibility, on the one hand, to independently tune the surface chemical composition that is given solely by the properties of the overcoat material and, on the other hand, the surface roughness that is primarily given by the size and number of NPs in the base layer, paves the way for the production of nanostructured materials with well-defined interfacial properties. This enabled production of surfaces with tailor-made wettability that may range from super-hydrophilic to super-hydrophobic (e.g., [60,266–273]), superwettable textiles for versatile oil/water separation with antibacterial properties [274] or nanostructured titanium coatings that mimic the surface roughness of a bone and, thus, facilitate the growth of osteoblasts [275]. Obviously, the sequence of deposition of NP and their subsequent overcoating may be repeated that allows for the fabrication of multiple stacks of particle and matrix layers [276]. For instance, in the recent study [277] multi-stack coatings composed of AgPt or AgAu NPs embedded in SiO_2 matrix were fabricated exhibiting reproducible diffusive memristive switching (see Figure 20).

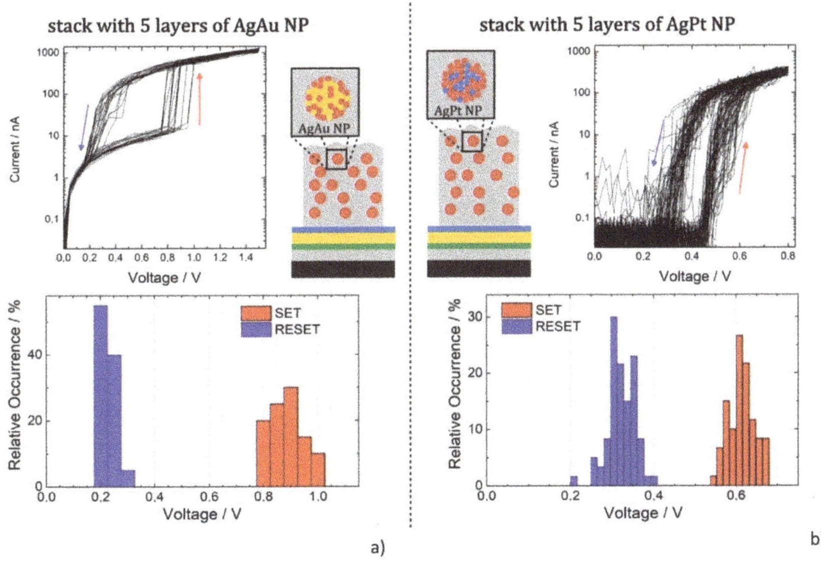

Figure 20. Multi-stack nanoparticle-based memristive devices relying on (**a**) AgAu and (**b**) AgPt nanoparticles exhibit diffusive memristive switching characteristics. For AgAu NP device, the switching characteristics are depicted for 20 consecutive cycles (top) with 2 nm SiO_2 separation layers. The corresponding histogram (bottom) shows a narrow distribution of SET (around 0.89 V) and RESET (around 0.23 V) voltages with a clear separation in between. For AgPt nanoparticle-based multi-stack device with SiO_2 separation layers of 4 nm each, the switching characteristics are depicted for 60 consecutive cycles (top) and the corresponding histogram (bottom) shows a narrow distribution of SET (around 0.61 V) and RESET (around 0.32 V) voltages with a clear separation in between. Reprinted from open-access source [277].

Finally, both the cluster beam and matrix deposition may be performed simultaneously [224,278,279]. This approach was among others successfully tested for the synthesis of Ag/a-C:H and Cu/a-C:H coatings using the deposition system schematically depicted in Figure 21a [280,281]. It consists of a magnetron-based gas aggregation source of metal NPs and a radio-frequency (RF) excitation electrode employed for the plasma polymerisation of n-hexane. The substrate to be coated was placed directly on the RF electrode and was positioned perpendicular to the beam of incoming NPs. This configuration enabled the production of metal/a-C:H nanocomposites (Figure 21b) with a variable number of embedded NPs that was controlled by the magnetron current and mechanical properties of a-C:H matrix regulated by the applied RF power. In a subsequent study [282] it was shown that under optimized conditions the produced Ag/a-C:H coatings deposited on Ti substrates provide excellent antibacterial performance against both Gram-negative (*Escherichia coli*) and Gram-positive (*Staphylococcus aureus*) bacteria and good biocompatibility, i.e., features needed in orthopaedics and other biomedical implants.

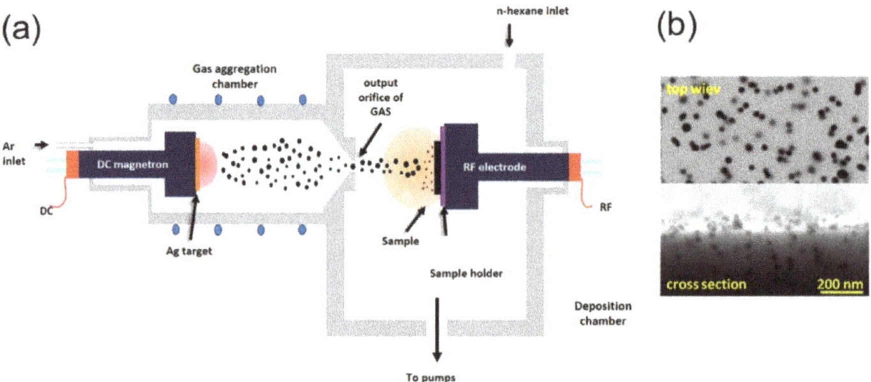

Figure 21. (**a**) Schematic representation of the experimental setup employed for the production of metal/a-C:H nanocomposites and (**b**) SEM images of the top view and cross-section of fabricated Ag/a-C:H coatings. Reprinted from [281] with permission from John Wiley and Sons.

4. Summary and Outlook

The main goal of this review was to overview and analyse the current state of the knowledge regarding the synthesis of nanomaterials using gas aggregation sources and outline their possible applications. As has been shown, the cluster beam deposition/implantation represents a very versatile way of preparing NP-assembled systems and coatings whose structure and functional properties can be tuned by beam parameters and deposition/implantation conditions (e.g., kinetic energy of NPs, their size, composition and structure, as well as properties of substrates or co-deposited material). Based on intensive experimental and theoretical studies of interactions of gas-aggregated NPs with various solid-state substrates, the basic mechanisms leading to the resulting structure of NP arrays/films and NP containing composites have been described and understood in the past two to three decades. This detailed knowledge subsequently paved the way for the rational and effective synthesis of novel NP-based materials with tailor-made structures that are applicable in numerous fields covering novel catalytic systems, magnetic media, platforms for sensing and detection, manufacturing of advanced electrical or electro-mechanical devices, including memristive materials or components for flexible electronics as well as biomedical tools. However, it is also important to note that, despite many advantages the cluster beam method offers, and the enormous development of the gas aggregation technique in past decades, the use of cluster sources at the industrial level is still relatively sparse.

There are several reasons for this paradoxical situation. First, the production capacity of NPs by low-pressure gas aggregation sources reaches only tens of mg per hour at present. This is relatively low, especially when compared with NP production techniques based on wet chemical methods, ball-milling, laser ablation in liquids [283–285], large-area atmospheric pressure glow discharges or plasma enhanced chemical vapour deposition [286–288] and utilization of the microplasma approach [289]. Therefore, the possible scale-up of the deposition systems and testing of new concepts of gas aggregation sources are nowadays under study intending to meet the requirements of industry on the quantity of synthesized NPs, areas that may be coated or the possibility to perform roll-to-roll deposition at sufficiently high speeds. The second important issue is related to the effectiveness of the NP production in terms of better utilisation of the source material, thus, reducing the costs and prolonging operation time of the cluster sources. In this filed, new technological solutions are urgently needed. Another aspect that may promote the wider utilisation of gas-phase synthesis of NPs is to make full use of the potential of this technique for the production of multi-component systems. This refers not only to the production of functional nanocomposites, in which the NPs are embedded in a hosting matrix, but also to the fabrication of multi-component NPs per se. Related to the latter, new highly promising strategies have been recently proposed and tested by different research groups. They utilize (i) a multi-magnetron approach, in which two or more sources of different materials are placed into one aggregation chamber; (ii) alloy or segmental multi-component sputtering targets; (iii) a combination of simultaneously running magnetron sputtering and plasma polymerization; or (iv) systems that are based on the in-flight coating/modification of NPs before their landing onto the substrate. However, the control of the structure and composition of multi-component NPs is still very challenging as the growth process is much more complex compared to mono-material particles. Due to this, the unavoidable prerequisite for the further development in this direction is the deeper understanding of the processes taking part during the growth and transport of NPs in the aggregation chamber or interaction of NPs with an auxiliary plasma used for their in-flight modification. This requires targeted experiments and novel numerical simulations whose complexity is at the edge of the computational capacity of current computers. Finally, novel classes of nanomaterials—nanofluids—with unique properties are nowadays discussed in the community [290,291]. In this case, the gas aggregation sources are believed to represent a powerful and effective means of production of such materials as the cluster beam deposition is, in principle, compatible with any liquid with low vapour pressure. This may open the way for the direct deposition of NPs into such liquids.

Author Contributions: Conceptualization: V.N.P.; writing—original draft preparation: V.N.P. and O.K.; writing—review and editing: V.N.P. and O.K.; visualization: V.N.P. and O.K. All authors have read and agreed to the published version of the manuscript.

Funding: This research received no external funding.

Conflicts of Interest: The authors declare no conflict of interest.

References

1. Barhoum, A.; Makhlouf, A.S.H. (Eds.) *Emerging Applications of Nanoparticles and Architecture Nanostructures*; Elsevier: Amsterdam, The Netherlands, 2018. [CrossRef]
2. Milani, P.; Sowwan, M. (Eds.) *Cluster Beam Deposition of Functional Nanomaterials and Devices*; Elsevier: Amsterdam, The Netherlands, 2020. [CrossRef]
3. Deepak, F.L. (Ed.) *Metal Nanoparticles and Clusters: Advances in Synthesis, Properties and Applications*; Springer: Cham, Switzerland, 2017. [CrossRef]
4. Ellis, P.R.; Brown, C.M.; Bishop, P.T.; Yin, J.; Cooke, K.; Terry, W.D.; Liu, J.; Yin, F.; Palmer, R.E. The cluster beam route to model catalysts and beyond. *Faraday Discuss.* **2016**, *188*, 39–56. [CrossRef] [PubMed]
5. Borghi, F.; Podestà, A.; Piazzoni, C.; Milani, P. Growth mechanism of cluster-assembled surfaces: From submonolayer to thin-film regime. *Phys. Rev. Appl.* **2018**, *9*, 044016. [CrossRef]
6. Grammatikopoulos, P.; Steinhauer, S.; Vernieres, J.; Singh, V.; Sowwan, M. Nanoparticle design by gas-phase synthesis. *Adv. Phys. X* **2016**, *1*, 81–100. [CrossRef]
7. Huttel, Y.; Martínez, L.; Mayoral, A.; Fernández, I. Gas-phase synthesis of nanoparticles: Present status and perspectives. *MRS Commun.* **2018**, *8*, 947–954. [CrossRef] [PubMed]

8. Haberland, H. Experimental methods. In *Clusters of Atoms and Molecules*; Haberland, H., Ed.; Springer: Berlin, Germany, 1994; pp. 207–252.

9. Pfund, A.H. Bismuth black and its applications. *Rev. Sci. Instrum.* **1930**, *1*, 397. [CrossRef]

10. Becker, E.W.; Bier, K.; Henkes, W. Strahlen aus kondensierten Atomen und Molekeln im Hochvakuum. *Eur. Phys. J. A* **1956**, *146*, 333–338. [CrossRef]

11. Henkes, W. Notizen: Ionisierung und Beschleunigung kondensierter Molekularstrahlen. *Z. Nat. A* **1961**, *16*, 842. [CrossRef]

12. Henkes, W. Massenspektrometrische Untersuchung von Strahlen aus kondensiertem Wasserstoff. *Z. Nat. A* **1962**, *17*, 786–789. [CrossRef]

13. Dole, M. Molecular beams of macroions. *J. Chem. Phys.* **1968**, *49*, 2240. [CrossRef]

14. Clampitt, R. Abstract: Intense field-emission ion source of liquid metals. *J. Vac. Sci. Technol.* **1975**, *12*, 1208. [CrossRef]

15. Helm, H. Field emission ion source of molecular cesium ions. *Rev. Sci. Instrum.* **1983**, *54*, 837. [CrossRef]

16. Dixon, A.; Ohana, R.; Sudraud, P.; Colliex, C.; Van De Walle, J. Field-ion emission from liquid tin. *Phys. Rev. Lett.* **1981**, *46*, 865–868. [CrossRef]

17. Hagena, O.F. Cluster formation in expanding supersonic jets: Effect of pressure, temperature, nozzle size, and test gas. *J. Chem. Phys.* **1972**, *56*, 1793. [CrossRef]

18. Hogg, E.O.; Silbernagel, B.G. Particle growth in flowing inert gases. *J. Appl. Phys.* **1974**, *45*, 593. [CrossRef]

19. Takagi, T.; Yamada, I.; Sasaki, A. An evaluation of metal and semiconductor films formed by ionized-cluster beam deposition. *Thin Solid Films* **1976**, *39*, 207–217. [CrossRef]

20. Bondybey, V.E. Laser induced fluorescence of metal clusters produced by laser vaporization: Gas phase spectrum of Pb2. *J. Chem. Phys.* **1981**, *74*, 6978. [CrossRef]

21. Dietz, T.G.; Duncan, M.A.; Powers, D.E.; Smalley, R.E. Laser production of supersonic metal cluster beams. *J. Chem. Phys.* **1981**, *74*, 6511–6512. [CrossRef]

22. Kroto, H.W.; Heath, J.R.; O'Brien, S.C.; Curl, R.F.; Smalley, R.E. C60: Buckminsterfullerene. *Nature* **1985**, *318*, 162–163. [CrossRef]

23. Begemann, W.; Meiwes-Broer, K.-H.; Lutz, H.O. Unimolecular decomposition of Sputtered Aln+, Cun+, and Sin+ clusters. *Phys. Rev. Lett.* **1986**, *56*, 2248–2251. [CrossRef]

24. Ganteför, G.; Siekmann, H.; Lutz, H.; Meiwes-Broer, K.-H. Pure metal and metal-doped rare-gas clusters grown in a pulsed ARC cluster ion source. *Chem. Phys. Lett.* **1990**, *165*, 293–296. [CrossRef]

25. Barborini, E.; Piseri, P.; Milani, P. A pulsed microplasma source of high intensity supersonic carbon cluster beams. *J. Phys. D Appl. Phys.* **1999**, *32*, L105–L109. [CrossRef]

26. Greene, J. Review article: Tracing the recorded history of thin-film sputter deposition: From the 1800s to 2017. *J. Vac. Sci. Technol. A* **2017**, *35*, 05C204. [CrossRef]

27. Haberland, H.; Karrais, M.; Mall, M. A new type of cluster and cluster ion source. *Eur. Phys. J. D* **1991**, *20*, 413–415. [CrossRef]

28. Polonskyi, O.; Ahadi, A.M.; Peter, T.; Fujioka, K.; Abraham, J.W.; Vasiliauskaite, E.; Hinz, A.; Strunskus, T.; Wolf, S.; Bonitz, M.; et al. Plasma based formation and deposition of metal and metal oxide nanoparticles using a gas aggregation source. *Eur. Phys. J. D* **2018**, *72*, 93. [CrossRef]

29. Shelemin, A.; Pleskunov, P.; Kousal, J.; Drewes, J.; Hanuš, J.; Ali-Ogly, S.; Nikitin, D.; Solar, P.; Kratochvíl, J.; Vaidulych, M.; et al. Nucleation and growth of magnetron-sputtered Ag nanoparticles as witnessed by time-resolved small angle X-ray scattering. *Part. Part. Syst. Charact.* **2019**, *37*, 1900436. [CrossRef]

30. Hagena, O.F. Cluster ion sources (invited). *Rev. Sci. Instrum.* **1992**, *63*, 2374–2379. [CrossRef]

31. Melinon, P.; Paillard, V.; Dupuis, V.; Perez, A.; Jensén, P.; Hoareau, A.; Pérez, J.; Tuaillon, J.; Broyer, M.; Vialle, J.; et al. From free clusters to cluster-assembled materials. *Int. J. Mod. Phys. B* **1995**, *9*, 339–397. [CrossRef]

32. Milani, P.; Iannotta, S. *Cluster Beam Synthesis of Nanostructured Materials*; Springer: Berlin, Germany, 1999.

33. Binns, C. Nanoclusters deposited on surfaces. *Surf. Sci. Rep.* **2001**, *44*, 1–49. [CrossRef]

34. Binns, C.; Trohidou, K.N.; Bansmann, J.; Baker, S.H.; A Blackman, J.; Bucher, J.-P.; Kechrakos, D.; Kleibert, A.; Louch, S.; Meiwes-Broer, K.-H.; et al. The behaviour of nanostructured magnetic materials produced by depositing gas-phase nanoparticles. *J. Phys. D Appl. Phys.* **2005**, *38*, R357–R379. [CrossRef]

35. Wegner, K.; Piseri, P.; Tafreshi, H.V.; Milani, P. Cluster beam deposition: A tool for nanoscale science and technology. *J. Phys. D Appl. Phys.* **2006**, *39*, R439–R459. [CrossRef]

36. Popok, V.N.; Campbell, E.E.B. Beams of atomic clusters: Effects on impact with solids. *Rev. Adv. Mater. Sci.* **2006**, *11*, 19–45.

37. Popok, V.N. Energetic cluster ion beams: Modification of surfaces and shallow layers. *Mater. Sci. Eng. R Rep.* **2011**, *72*, 137–157. [CrossRef]

38. Yamada, I. Historical milestones and future prospects of cluster ion beam technology. *Appl. Surf. Sci.* **2014**, *310*, 77–88. [CrossRef]

39. Palmer, R.E.; Cao, L.; Yin, F. Note: Proof of principle of a new type of cluster beam source with potential for scale-up. *Rev. Sci. Instrum.* **2016**, *87*, 46103. [CrossRef] [PubMed]

40. Cai, R.; Cao, L.; Griffin, R.; Chansai, S.; Hardacre, C.; Palmer, R.E. Scale-up of cluster beam deposition to the gram scale with the matrix assembly cluster source for heterogeneous catalysis (propylene combustion). *AIP Adv.* **2020**, *10*, 025314. [CrossRef]

41. Shelemin, A.; Kylián, O.; Hanuš, J.; Choukourov, A.; Melnichuk, I.; Serov, A.; Slavinska, D.; Biederman, H. Preparation of metal oxide nanoparticles by gas aggregation cluster source. *Vacuum* **2015**, *120*, 162–169. [CrossRef]

42. Martínez, L.; Lauwaet, K.; Santoro, G.; Sobrado, J.M.; Peláez, R.J.; Herrero, V.J.; Tanarro, I.; Ellis, G.; Cernicharo, J.; Joblin, C.; et al. Precisely controlled fabrication, manipulation and In-Situ analysis of Cu based nanoparticles. *Sci. Rep.* **2018**, *8*, 7250. [CrossRef]

43. Popok, V.N.; Jeppesen, C.M.; Fojan, P.; Kuzminova, A.; Hanuš, J.; Kylián, O. Comparative study of antibacterial properties of polystyrene films with TiO x and Cu nanoparticles fabricated using cluster beam technique. *Beilstein J. Nanotechnol.* **2018**, *9*, 861–869. [CrossRef]

44. Vahl, A.; Strobel, J.; Reichstein, W.; Polonskyi, O.; Strunskus, T.; Kienle, L.; Faupel, F. Single target sputter deposition of alloy nanoparticles with adjustable composition via a gas aggregation cluster source. *Nanotechnology* **2017**, *28*, 175703. [CrossRef]

45. Ayesh, A.I. Size-selected fabrication of alloy nanoclusters by plasma-gas condensation. *J. Alloys Compd.* **2018**, *745*, 299–305. [CrossRef]

46. Xu, Y.-H.; Wang, J.-P. Direct gas-phase synthesis of heterostructured nanoparticles through phase separation and surface segregation. *Adv. Mater.* **2008**, *20*, 994–999. [CrossRef]

47. Martínez, L.; Mayoral, A.; Espiñeira, M.; Roman, E.; Palomares, F.J.; Huttel, Y. Core@shell, Au@TiOx nanoparticles by gas phase synthesis. *Nanoscale* **2017**, *9*, 6463–6470. [CrossRef] [PubMed]

48. Llamosa, D.; Ruano, M.; Martinez, L.; Mayoral, A.; Roman, E.; Garcia-Hernandez, M.; Huttel, Y. The ultimate step towards a tailored engineering of core@shell and core@shell@shell nanoparticles. *Nanoscale* **2014**, *6*, 13483–13486. [CrossRef] [PubMed]

49. Hanuš, J.; Vaidulych, M.; Kylián, O.; Choukourov, A.; Kousal, J.; Khalakhan, I.; Cieslar, M.; Solar, P.; Biederman, H. Fabrication of Ni@Ti core–shell nanoparticles by modified gas aggregation source. *J. Phys. D Appl. Phys.* **2017**, *50*, 475307. [CrossRef]

50. Yin, F.; Wang, Z.-W.; Palmer, R.E. Controlled formation of mass-selected Cu–Au core–shell cluster beams. *J. Am. Chem. Soc.* **2011**, *133*, 10325–10327. [CrossRef]

51. Singh, V.; Cassidy, C.; Grammatikopoulos, P.; Djurabekova, F.; Nordlund, K.; Sowwan, M. Heterogeneous gas-phase synthesis and molecular dynamics modeling of janus and core–satellite Si–Ag nanoparticles. *J. Phys. Chem. C* **2014**, *118*, 13869–13875. [CrossRef]

52. Solar, P.; Hanuš, J.; Cieslar, M.; Košutová, T.; Škorvánková, K.; Kylián, O.; Kúš, P.; Biederman, H. Composite Ni@Ti nanoparticles produced in arrow-shaped gas aggregation source. *J. Phys. D Appl. Phys.* **2020**, *53*, 195303. [CrossRef]

53. Solar, P.; Polonskyi, O.; Olbricht, A.; Hinz, A.; Shmielen, A.; Kylián, O.; Choukourov, A.; Faupel, F.; Biederman, H. Single-step generation of metal-plasma polymer multicore@shell nanoparticles from the gas phase. *Sci. Rep.* **2017**, *7*, 8514. [CrossRef]

54. Kylián, O.; Shelemin, A.; Solar, P.; Pleskunov, P.; Nikitin, D.; Kuzminova, A.; Štefaníková, R.; Kúš, P.; Cieslar, M.; Hanuš, J.; et al. Magnetron sputtering of polymeric targets: From thin films to heterogeneous metal/plasma polymer nanoparticles. *Materials* **2019**, *12*, 2366. [CrossRef]

55. Kylián, O.; Kuzminova, A.; Štefaníková, R.; Hanuš, J.; Solař, P.; Kúš, P.; Cieslar, M.; Choukourov, A.; Biederman, H. Silver/plasma polymer strawberry-like nanoparticles produced by gas-phase synthesis. *Mater. Lett.* **2019**, *253*, 238–241. [CrossRef]

56. Pratontep, S.; Carroll, S.J.; Xirouchaki, C.; Streun, M.; Palmer, R.E. Size-selected cluster beam source based on radio frequency magnetron plasma sputtering and gas condensation. *Rev. Sci. Instrum.* **2005**, *76*, 45103. [CrossRef]

57. Popok, V.N.; Barke, I.; Campbell, E.E.B.; Meiwes-Broer, K.-H. Cluster–surface interaction: From soft landing to implantation. *Surf. Sci. Rep.* **2011**, *66*, 347–377. [CrossRef]

58. Kappes, M.M.; Leutwyler, S. Molecular beams of clusters. In *Atomic and Molecular Beam Methods*; Scoles, G., Ed.; Oxford University Press: New York, NY, USA, 1988; Volume 1, pp. 380–415.

59. De Heer, W.A. The physics of simple metal clusters: Experimental aspects and simple models. *Rev. Mod. Phys.* **1993**, *65*, 611–676. [CrossRef]

60. Petr, M.; Kylián, O.; Hanuš, J.; Kuzminova, A.; Vaidulych, M.; Khalakhan, I.; Choukourov, A.; Slavinska, D.; Biederman, H. Surfaces with roughness gradient and invariant surface chemistry produced by means of gas aggregation source and magnetron sputtering. *Plasma Process. Polym.* **2016**, *13*, 663–671. [CrossRef]

61. Caruso, F.; Bellacicca, A.; Milani, P. High-throughput shadow mask printing of passive electrical components on paper by supersonic cluster beam deposition. *Appl. Phys. Lett.* **2016**, *108*, 163501. [CrossRef]

62. Kratochvil, J.; Stranak, V.; Kousal, J.; Kus, P.; Kylian, O. Theoretical and experimental analysis of defined 2D-graded two-metal nanoparticle-build surfaces. *Appl. Surf. Sci.* **2020**, *511*, 145530. [CrossRef]

63. Meakin, P.; Ramanlal, P.; Sander, L.M.; Ball, R.C. Ballistic deposition on surfaces. *Phys. Rev. A* **1986**, *34*, 5091–5103. [CrossRef]

64. Vold, M.J. Computer simulation of floc formation in a colloidal suspension. *J. Colloid Sci.* **1963**, *18*, 684–695. [CrossRef]

65. Sutherland, D.N. A theoretical model of floc structure. *J. Colloid Interface Sci.* **1967**, *25*, 373–380. [CrossRef]

66. Barabasi, A.-L.; Stanley, H.E. *Fractal Concepts in Surface Growth*; Cambridge University Press: New York, NY, USA, 1995. [CrossRef]

67. Family, F.; Vicsek, T. (Eds.) *Dynamics of Fractal Surfaces*; World Scientific: Singapore, 1991. [CrossRef]

68. Konstandopoulos, A.G. Deposit growth dynamics: Particle sticking and scattering phenomena. *Powder Technol.* **2000**, *109*, 262–277. [CrossRef]

69. Podestà, A.; Borghi, F.; Indrieri, M.; Bovio, S.; Piazzoni, C.; Milani, P. Nanomanufacturing of titania interfaces with controlled structural and functional properties by supersonic cluster beam deposition. *J. Appl. Phys.* **2015**, *118*, 234309. [CrossRef]

70. Bardotti, L.; Jensén, P.; Hoareau, A.; Treilleux, M.; Cabaud, B.; Perez, A.; Aires, F.C.S. Diffusion and aggregation of large antimony and gold clusters deposited on graphite. *Surf. Sci.* **1996**, *367*, 276–292. [CrossRef]

71. Bardotti, L.; Jensen, P.; Hoareau, A.; Treilleux, M.; Cabaud, B. Experimental observation of fast diffusion of large antimony clusters on graphite surfaces. *Phys. Rev. Lett.* **1995**, *74*, 4694–4697. [CrossRef]

72. Deltour, P.; Jensen, P.; Barrat, J.-L. Fast diffusion of a lennard-jones cluster on a crystalline surface. *Phys. Rev. Lett.* **1997**, *78*, 4597–4600. [CrossRef]

73. Ryu, J.H.; Seo, D.; Kim, D.-H.; Lee, H.M. Molecular dynamics simulations of the diffusion and rotation of Pt nanoclusters supported on graphite. *Phys. Chem. Chem. Phys.* **2009**, *11*, 503–507. [CrossRef] [PubMed]

74. Perez, A.; Melinon, P.; Dupuis, V.; Jensen, P.; Prével, B.; Tuaillon, J.; Bardotti, L.; Martet, C.; Treilleux, M.; Broyer, M.; et al. Cluster assembled materials: A novel class of nanostructured solids with original structures and properties. *J. Phys. D Appl. Phys.* **1997**, *30*, 709–721. [CrossRef]

75. Grammatikopoulos, P.; Sowwan, M.; Kioseoglou, J. Computational modeling of nanoparticle coalescence. *Adv. Theory Simul.* **2019**, *2*, 1900013. [CrossRef]

76. Lehtinen, K.E.; Zachariah, M. Energy accumulation in nanoparticle collision and coalescence processes. *J. Aerosol Sci.* **2002**, *33*, 357–368. [CrossRef]

77. Lehtinen, K.E.J.; Zachariah, M.R. Effect of coalescence energy release on the temporal shape evolution of nanoparticles. *Phys. Rev. B* **2001**, *63*, 205402. [CrossRef]

78. Singh, V.; Grammatikopoulos, P.; Cassidy, C.; Benelmekki, M.; Bohra, M.; Hawash, Z.; Baughman, K.W.; Sowwan, M. Assembly of tantalum porous films with graded oxidation profile from size-selected nanoparticles. *J. Nanoparticle Res.* **2014**, *16*, 2373. [CrossRef]

79. Yoon, B.; Akulin, V.; Cahuzac, P.; Carlier, F.; De Frutos, M.; Masson, A.; Mory, C.; Colliex, C.; Bréchignac, C. Morphology control of the supported islands grown from soft-landed clusters. *Surf. Sci.* **1999**, *443*, 76–88. [CrossRef]

80. Carroll, S.J.; Seeger, K.; Palmer, R.E. Trapping of size-selected Ag clusters at surface steps. *Appl. Phys. Lett.* **1998**, *72*, 305–307. [CrossRef]

81. Jensen, P. Growth of nanostructures by cluster deposition: Experiments and simple models. *Rev. Mod. Phys.* **1999**, *71*, 1695–1735. [CrossRef]

82. Jensen, P.; Clément, A.; Lewis, L.J. Diffusion of nanoclusters on non-ideal surfaces. *Phys. E Low-Dimens. Syst. Nanostruct.* **2004**, *21*, 71–76. [CrossRef]

83. Jensen, P.; Clément, A.; Lewis, L.J. Diffusion of nanoclusters. *Comput. Mater. Sci.* **2004**, *30*, 137–142. [CrossRef]

84. Schmidt, M.; Kébaïli, N.; Lando, A.; Benrezzak, S.; Baraton, L.; Cahuzac, P.; Masson, A.; Bréchignac, C. Bent graphite surfaces as guides for cluster diffusion and anisotropic growth. *Phys. Rev. B* **2008**, *77*, 205420. [CrossRef]

85. Pérez, A.; Bardotti, L.; Prével, B.; Jensen, P.; Treilleux, M.; Melinon, P.; Gierak, J.; Faini, G.; Mailly, D. Quantum-dot systems prepared by 2D organization of nanoclusters preformed in the gas phase on functionalized substrates. *New J. Phys.* **2002**, *4*, 76. [CrossRef]

86. Bardotti, L.; Prevel, B.; Jensen, P.; Treilleux, M.; Melinon, P.; Pérez, A.; Gierak, J.; Faini, G.; Mailly, D. Organizing nanoclusters on functionalized surfaces. *Appl. Surf. Sci.* **2002**, *191*, 205–210. [CrossRef]

87. Prevel, B.; Bardotti, L.; Fanget, S.; Hannour, A.; Melinon, P.; Pérez, A.; Gierak, J.; Faini, G.; Bourhis, E.; Mailly, D. Gold nanoparticle arrays on graphite surfaces. *Appl. Surf. Sci.* **2004**, *226*, 173–177. [CrossRef]

88. Melinon, P.; Hannour, A.; Prével, B.; Bardotti, L.; Bernstein, E.; Perez, A.; Gierak, J.; Bourhis, E.; Mailly, D. Functionalizing surfaces with arrays of clusters: Role of the defects. *J. Cryst. Growth* **2005**, *275*, 317–324. [CrossRef]

89. Novikov, S.M.; Popok, V.N.; Fiutowski, J.; Arsenin, A.V.; Volkov, V.S. Plasmonic properties of nanostructured graphene with silver nanoparticles. *J. Phys. Conf. Ser.* **2020**, *1461*, 012119. [CrossRef]

90. Mondal, S.; Chowdhury, D. Controlled deposition of size-selected metal nanoclusters on prepatterned substrate. *Surf. Coat. Technol.* **2020**, *393*, 125776. [CrossRef]

91. Kébaili, N.; Benrezzak, S.; Cahuzac, P.; Masson, A.; Bréchignac, C. Diffusion of silver nanoparticles on carbonaceous materials. Cluster mobility as a probe for surface characterization. *Eur. Phys. J. D* **2009**, *52*, 115–118. [CrossRef]

92. Bardotti, L.; Tournus, F.; Melinon, P.; Pellarin, M.; Broyer, M. Mass-selected clusters deposited on graphite: Spontaneous organization controlled by cluster surface reaction. *Phys. Rev. B* **2011**, *83*, 035425. [CrossRef]

93. Alayan, R.; Arnaud, L.; Broyer, M.; Cottancin, E.; Lermé, J.; Marhaba, S.; Vialle, J.L.; Pellarin, M. Organization of size-selected platinum and indium clusters soft-landed on surfaces. *Phys. Rev. B* **2007**, *76*, 075424. [CrossRef]

94. Tainoff, D.; Bardotti, L.; Tournus, F.; Guiraud, G.; Boisron, O.; Melinon, P. Self-organization of size-selected bare platinum nanoclusters: Toward ultra-dense catalytic systems. *J. Phys. Chem. C* **2008**, *112*, 6842–6849. [CrossRef]

95. Tournus, F.; Bardotti, L.; Dupuis, V. Size-dependent morphology of CoPt cluster films on graphite: A route to self-organization. *J. Appl. Phys.* **2011**, *109*, 114309. [CrossRef]

96. Hou, Q.; Hou, M.; Bardotti, L.; Prével, B.; Melinon, P.; Perez, A. Deposition of AuN clusters on Au(111) surfaces. I. Atomic-scale modeling. *Phys. Rev. B* **2000**, *62*, 2825–2834. [CrossRef]

97. Jimenez-Saez, J.; Pérez-Martin, C.; Jiménez-Rodríguez, J. A molecular dynamics study of atomic rearrangements in Cu clusters softly deposited on an Au(001) surface. *Nucl. Instrum. Methods Phys. Res. B* **2006**, *249*, 816–819. [CrossRef]

98. Järvi, T.T.; Kuronen, A.; Meinander, K.; Nordlund, K.; Albe, K. Contact epitaxy by deposition of Cu, Ag, Au, Pt, and Ni nanoclusters on (100) surfaces: Size limits and mechanisms. *Phys. Rev. B* **2007**, *75*, 115422. [CrossRef]

99. Meinander, K.; Frantz, J.; Nordlund, K.; Keinonen, J. Upper size limit of complete contact epitaxy. *Thin Solid Films* **2003**, *425*, 297–303. [CrossRef]

100. Reichel, R.; Partridge, J.G.; Natali, F.; Matthewson, T.; Brown, S.A.; Lassesson, A.; MacKenzie, D.M.A.; Ayesh, A.I.; Tee, K.C.; Awasthi, A.; et al. From the adhesion of atomic clusters to the fabrication of nanodevices. *Appl. Phys. Lett.* **2006**, *89*, 213105. [CrossRef]

101. Awasthi, A.; Hendy, S.C.; Zoontjens, P.; Brown, S.A.; Natali, F. Molecular dynamics simulations of reflection and adhesion behavior in Lennard-Jones cluster deposition. *Phys. Rev. B* **2007**, *76*, 115437. [CrossRef]

102. Reichel, R.; Partridge, J.G.; Brown, S.A. Characterization of a template process for conducting cluster-assembled wires. *Appl. Phys. A* **2009**, *97*, 315–321. [CrossRef]

103. Carroll, S.J.; Weibel, P.; Von Issendorff, B.; Kuipers, L.; Palmer, R.E. The impact of size-selected Ag clusters on graphite: An STM study. *J. Phys. Condens. Matter* **1996**, *8*, L617–L624. [CrossRef]

104. Di Vece, M.; Palomba, S.; E Palmer, R. Pinning of size-selected gold and nickel nanoclusters on graphite. *Phys. Rev. B* **2005**, *72*, 073407. [CrossRef]

105. Gibilisco, S.; Di Vece, M.; Palomba, S.; Faraci, G.; Palmer, R.E. Pinning of size-selected Pd nanoclusters on graphite. *J. Chem. Phys.* **2006**, *125*, 84704. [CrossRef]

106. Vučković, S.; Svanqvist, M.; Popok, V.N. Laser ablation source for formation and deposition of size-selected metal clusters. *Rev. Sci. Instrum.* **2008**, *79*, 73303. [CrossRef]

107. Vučković, S.; Samela, J.; Nordlund, K.; Popok, V.N. Pinning of size-selected Co clusters on highly ordered pyrolytic graphite. *Eur. Phys. J. D* **2009**, *52*, 107–110. [CrossRef]

108. Popok, V.N.; Vučković, S.; Samela, J.; Järvi, T.T.; Nordlund, K.; Campbell, E.E.B. Stopping of energetic cobalt clusters and formation of radiation damage in graphite. *Phys. Rev. B* **2009**, *80*, 205419. [CrossRef]

109. Smith, R.; Nock, C.; Kenny, S.; BelBruno, J.J.; Di Vece, M.; Palomba, S.; Palmer, R.E. Modeling the pinning of Au and Ni clusters on graphite. *Phys. Rev. B* **2006**, *73*, 125429. [CrossRef]

110. Smith, R.; Kenny, S.; BelBruno, J.; Palmer, R. Chapter 15 Modelling the structure and dynamics of metal nanoclusters deposited on graphite. In *Atomic Clusters: From Gas Phase to Deposition*; Woodruff, D.P., Ed.; Elsevier: Amsterdam, The Netherlands, 2007; pp. 589–616.

111. Convers, P.; Monot, R.; Harbich, W.; Seminara, L. Implantation of size-selected silver clusters into graphite. *Eur. Phys. J. D* **2004**, *29*, 49–56. [CrossRef]

112. Samela, J.; Nordlund, K.; Keinonen, J.; Popok, V.N.; Campbell, E.E.B. Argon cluster impacts on layered silicon, silica, and graphite surfaces. *Eur. Phys. J. D* **2007**, *43*, 181–184. [CrossRef]

113. Chang, H.; Bard, A.J. Scanning tunneling microscopy studies of carbon-oxygen reactions on highly oriented pyrolytic graphite. *J. Am. Chem. Soc.* **1991**, *113*, 5588–5596. [CrossRef]

114. Datta, S.S.; Strachan, D.R.; Khamis, S.M.; Johnson, A.T.C. Crystallographic etching of few-layer graphene. *Nano Lett.* **2008**, *8*, 1912–1915. [CrossRef] [PubMed]

115. Severin, N.; Kirstein, S.; Sokolov, I.M.; Rabe, J.P. Rapid trench channeling of graphenes with catalytic silver nanoparticles. *Nano Lett.* **2009**, *9*, 457–461. [CrossRef] [PubMed]

116. Sigmund, P. Mechanisms and theory of physical sputtering by particle impact. *Nucl. Instum. Meth. Phys. Res. B* **1987**, *27*, 1–20. [CrossRef]

117. Nastasi, M.; Mayer, J.W. *Ion Implantation and Synthesis of Materials*; Springer: Berlin, Germany, 2006. [CrossRef]

118. Wilson, I.H.; Zheng, N.J.; Knipping, U.; Tsong, I.S.T. Effects of isolated atomic collision cascades on SiO2/Si interfaces studied by scanning tunneling microscopy. *Phys. Rev. B* **1988**, *38*, 8444–8450. [CrossRef]

119. Neumann, R. Scanning probe microscopy of ion-irradiated materials. *Nucl. Instrum. Methods Phys. Res. B* **1999**, *151*, 42–55. [CrossRef]

120. Colla, T.J.; Aderjan, R.; Kissel, R.; Urbassek, H.M. Sputtering of Au (111) induced by 16-keV Au cluster bombardment: Spikes, craters, late emission, and fluctuations. *Phys. Rev. B* **2000**, *62*, 8487–8493. [CrossRef]

121. Bräuchle, G.; Richard-Schneider, S.; Illig, D.; Rockenberger, J.; Beck, R.D.; Kappes, M.M. Etching nanometer sized holes of variable depth from carbon cluster impact induced defects on graphite surfaces. *Appl. Phys. Lett.* **1995**, *67*, 52–54. [CrossRef]

122. Seki, T.; Aoki, T.; Tanomura, M.; Matsuo, J.; Yamada, I. Energy dependence of a single trace created by C60 ion impact. *Mater. Chem. Phys.* **1998**, *54*, 143–146. [CrossRef]

123. Allen, L.P.; Insepov, Z.; Fenner, D.B.; Santeufemio, C.; Brooks, W.; Jones, K.S.; Yamada, I. Craters on silicon surfaces created by gas cluster ion impacts. *J. Appl. Phys.* **2002**, *92*, 3671–3678. [CrossRef]

124. Popok, V.N.; Prasalovich, S.; Campbell, E.E.B. Complex crater formation on silicon surfaces by low-energy Arn+ cluster ion implantation. *Surf. Sci.* **2004**, *566*, 1179–1184. [CrossRef]

125. Popok, V.N.; Prasalovich, S.; Campbell, E.E.B. Surface nanostructuring by implantation of cluster ions. *Vacuum* **2004**, *76*, 265–272. [CrossRef]

126. Samela, J.; Nordlund, K.; Keinonen, J.; Popok, V.N. Comparison of silicon potentials for cluster bombardment simulations. *Nucl. Instrum. Methods Phys. Res. B* **2007**, *255*, 253–258. [CrossRef]

127. Samela, J.; Nordlund, K.; Popok, V.N.; Campbell, E.E.B. Origin of complex impact craters on native oxide coated silicon surfaces. *Phys. Rev. B* **2008**, *77*, 075309. [CrossRef]

128. Prasalovich, S.; Popok, V.N.; Persson, P.O.; Campbell, E.E.B. Experimental studies of complex crater formation under cluster implantation of solids. *Eur. Phys. J. D* **2005**, *36*, 79–88. [CrossRef]

129. Popok, V.N.; Jensen, J.; Vuckovic, S.; Macková, A.; Trautmann, C. Formation of surface nanostructures on rutile (TiO$_2$): Comparative study of low-energy cluster ion and high-energy monoatomic ion impact. *J. Phys. D Appl. Phys.* **2009**, *42*, 205303. [CrossRef]

130. Yamaguchi, Y.; Gspann, J. Large-scale molecular dynamics simulations of cluster impact and erosion processes on a diamond surface. *Phys. Rev. B* **2002**, *66*, 155408. [CrossRef]

131. Djurabekova, F.; Samela, J.; Timko, H.; Nordlund, K.; Calatroni, S.; Taborelli, M.; Wuensch, W. Crater formation by single ions, cluster ions and ion showers. *Nucl. Instrum. Methods Phys. Res. B* **2012**, *272*, 374–376. [CrossRef]

132. Popok, V.N.; Samela, J.; Nordlund, K.; Popov, V.P. Impact of keV-energy argon clusters on diamond and graphite. *Nucl. Instrum. Methods Phys. Res. B* **2012**, *282*, 112–115. [CrossRef]

133. Popok, V.N.; Samela, J.; Nordlund, K.; Campbell, E.E.B. Stopping of energetic argon cluster ions in graphite: Role of cluster momentum and charge. *Phys. Rev. B* **2010**, *82*, 201403. [CrossRef]

134. Gruber, A. Nanoparticle impact micromachining. *J. Vac. Sci. Technol. B* **1997**, *15*, 2362. [CrossRef]

135. Popok, V.N.; Prasalovich, S.; Campbell, E.E.B. Nanohillock formation by impact of small low-energy clusters with surfaces. *Nucl. Instrum. Methods Phys. Res. B* **2003**, *207*, 145–153. [CrossRef]

136. Cleveland, C.L.; Landman, U. Dynamics of cluster-surface collisions. *Science* **1992**, *257*, 355–361. [CrossRef]

137. Melosh, H.J.; Ivanov, B.A. Impact crater collapse. *Annu. Rev. Earth Planet. Sci.* **1999**, *27*, 385–415. [CrossRef]

138. Yamada, I. Materials processing by gas cluster ion beams. *Mater. Sci. Eng. R Rep.* **2001**, *34*, 231–295. [CrossRef]

139. Nazarov, A.V.; Chernysh, V.; Nordlund, K.; Djurabekova, F.; Zhao, J. Spatial distribution of particles sputtered from single crystals by gas cluster ions. *Nucl. Instrum. Methods Phys. Res. B* **2017**, *406*, 518–522. [CrossRef]

140. Matsuo, J.; Toyoda, N.; Akizuki, M.; Yamada, I. Sputtering of elemental metals by Ar cluster ions. *Nucl. Instrum. Methods Phys. Res. B* **1997**, *121*, 459–463. [CrossRef]

141. Toyoda, N.; Yamada, I. Size effects of gas cluster ions on beam transport, amorphous layer formation and sputtering. *Nucl. Instrum. Methods Phys. Res. B* **2009**, *267*, 1415–1419. [CrossRef]

142. Winograd, N. The magic of cluster SIMS. *Anal. Chem.* **2005**, *77*, 142 A–149 A. [CrossRef]

143. Gilmore, I. SIMS of organics—Advances in 2D and 3D imaging and future outlook. *J. Vac. Sci. Technol. A* **2013**, *31*, 050819. [CrossRef]

144. Shard, A.G.; Havelund, R.; Seah, M.P.; Spencer, S.J.; Gilmore, I.; Winograd, N.; Mao, D.; Miyayama, T.; Niehuis, E.; Rading, D.; et al. Argon cluster ion beams for organic depth profiling: Results from a VAMAS interlaboratory study. *Anal. Chem.* **2012**, *84*, 7865–7873. [CrossRef]

145. Seah, M.P.; Gilmore, I. Cluster primary ion sputtering: Correlations in secondary ion intensities in TOF SIMS. *Surf. Interface Anal.* **2010**, *43*, 228–235. [CrossRef]

146. Angerer, T.B.; Blenkinsopp, P.; Fletcher, J.S. High energy gas cluster ions for organic and biological analysis by time-of-flight secondary ion mass spectrometry. *Int. J. Mass Spectrom.* **2015**, *377*, 591–598. [CrossRef]

147. Paruch, R.J.; Garrison, B.J.; Mlynek, M.; Postawa, Z. On universality in sputtering yields due to cluster bombardment. *J. Phys. Chem. Lett.* **2014**, *5*, 3227–3230. [CrossRef]

148. Averback, R.; Ghaly, M. MD studies of the interactions of low energy particles and clusters with surfaces. *Nucl. Instrum. Methods Phys. Res. B* **1994**, *90*, 191–201. [CrossRef]

149. Yamamura, Y. Sputtering by cluster ions. *Nucl. Instrum. Methods Phys. Res. B* **1988**, *33*, 493–496. [CrossRef]

150. Shulga, V.; Sigmund, P. Penetration of slow gold clusters through silicon. *Nucl. Instrum. Methods Phys. Res. B* **1990**, *47*, 236–242. [CrossRef]

151. Shulga, V.I.; Vicanek, M.; Sigmund, P. Pronounced nonlinear behavior of atomic collision sequences induced by keV-energy heavy ions in solids and molecules. *Phys. Rev. A* **1989**, *39*, 3360–3372. [CrossRef] [PubMed]

152. Ravagnan, L.; Divitini, G.; Rebasti, S.; Marelli, M.; Piseri, P.; Milani, P. Poly(methyl methacrylate)-palladium clusters nanocomposite formation by supersonic cluster beam deposition: A method for microstructured metallization of polymer surfaces. *J. Phys. D Appl. Phys.* **2009**, *42*, 082002. [CrossRef]

153. Cardia, R.; Melis, C.; Colombo, L. Neutral-cluster implantation in polymers by computer experiments. *J. Appl. Phys.* **2013**, *113*, 224307. [CrossRef]

154. Ziegler, J.F.; Ziegler, M.; Biersack, J. SRIM—The stopping and range of ions in matter (2010). *Nucl. Instrum. Methods Phys. Res. B* **2010**, *268*, 1818–1823. [CrossRef]

155. Möller, W.; Eckstein, W.; Biersack, J. Tridyn-binary collision simulation of atomic collisions and dynamic composition changes in solids. *Comput. Phys. Commun.* **1988**, *51*, 355–368. [CrossRef]

156. Toyoda, N.; Yamada, I. Gas cluster ion beam equipment and applications for surface processing. *IEEE Trans. Plasma Sci.* **2008**, *36*, 1471–1488. [CrossRef]

157. Gilmer, G.; Roland, C.; Stock, D.; Jaraíz, M.; De La Rubia, T.D. Simulations of thin film deposition from atomic and cluster beams. *Mater. Sci. Eng. B* **1996**, *37*, 1–7. [CrossRef]

158. Anders, C.; Urbassek, H.M. Cluster-size dependence of ranges of 100eV/atom Aun clusters. *Nucl. Instrum. Methods Phys. Res. B* **2005**, *228*, 57–63. [CrossRef]

159. Pratontep, S.; Preece, P.; Xirouchaki, C.; Palmer, R.E.; Sanz-Navarro, C.; Kenny, S.; Smith, R. Scaling relations for implantation of size-selected Au, Ag, and Si clusters into graphite. *Phys. Rev. Lett.* **2003**, *90*, 55503. [CrossRef]

160. Popok, V.N.; Samela, J.; Nordlund, K.; Popov, V.P. Implantation of keV-energy argon clusters and radiation damage in diamond. *Phys. Rev. B* **2012**, *85*, 033405. [CrossRef]

161. Tyo, E.C.; Vajda, S. Catalysis by clusters with precise numbers of atoms. *Nat. Nanotechnol.* **2015**, *10*, 577–588. [CrossRef] [PubMed]

162. Heiz, U.; Bullock, E.L. Fundamental aspects of catalysis on supported metal clusters. *J. Mater. Chem.* **2004**, *14*, 564–577. [CrossRef]

163. Bansmann, J.; Baker, S.; Binns, C.; Blackman, J.; Bucher, J.-P.; Dorantesdavila, J.; Dupuis, V.; Favre, L.; Kechrakos, D.; Kleibert, A. Magnetic and structural properties of isolated and assembled clusters. *Surf. Sci. Rep.* **2005**, *56*, 189–275. [CrossRef]

164. Benel, C.; Reisinger, T.; Kruk, R.; Hahn, H. Cluster-assembled nanocomposites: Functional properties by design. *Adv. Mater.* **2018**, *31*, 1806634. [CrossRef] [PubMed]

165. Kreibig, U.; Volmer, M. Optics of nanosized metals. In *Handbook of Optical Properties: Optics of Small Particles, Interfaces, and Surfaces*; Hummel, R.E., Wissman, P., Eds.; CRC Press: Boca Raton, FL, USA, 1997; Volume 2, pp. 145–190.

166. Pelton, M.; Aizpurua, J.; Bryant, G. Metal-nanoparticle plasmonics. *Laser Photon. Rev.* **2008**, *2*, 136–159. [CrossRef]

167. Halas, N.J.; Lal, S.; Chang, W.-S.; Link, S.; Nordlander, P. Plasmons in strongly coupled metallic nanostructures. *Chem. Rev.* **2011**, *111*, 3913–3961. [CrossRef] [PubMed]

168. Jeon, T.Y.; Park, S.-G.; Kim, S.-H.; Kim, D.J.; Kim, D.-H. Nanostructured plasmonic substrates for use as SERS sensors. *Nano Converg.* **2016**, *3*, 2957. [CrossRef]

169. Ghisleri, C.; Siano, M.; Ravagnan, L.; Potenza, M.A.C.; Milani, P. Nanocomposite-based stretchable optics. *Laser Photon. Rev.* **2013**, *7*, 1020–1026. [CrossRef]

170. Marom, S.; Dorresteijn, M.; Modi, R.; Podesta, A.; Di Vece, M. Silver nanoparticles from a gas aggregation nanoparticle source for plasmonic efficiency enhancement in a-Si solar cells. *Mater. Res. Express* **2019**, *6*, 045012. [CrossRef]

171. Minnai, C.; Bellacicca, A.; Brown, S.A.; Milani, P. Facile fabrication of complex networks of memristive devices. *Sci. Rep.* **2017**, *7*, 7955. [CrossRef]

172. Santaniello, T.; Milani, P. Additive nano-manufacturing of 3D printed electronics using supersonic cluster beam deposition. *Front. Nanosci.* **2020**, *15*, 313–333. [CrossRef]

173. Santaniello, T.; Migliorini, L.; Yan, Y.; Lenardi, C.; Milani, P. Supersonic cluster beam fabrication of metal–ionogel nanocomposites for soft robotics. *J. Nanopart. Res.* **2018**, *20*, 250. [CrossRef]

174. Fostner, S.; Nande, A.; Smith, A.; Gazoni, R.M.; Grigg, J.; Temst, K.; Van Bael, M.J.; Brown, S.A. Percolating transport in superconducting nanoparticle films. *J. Appl. Phys.* **2017**, *122*, 223905. [CrossRef]

175. Zheng, M.; Li, W.; Xu, M.; Xu, N.; Chen, P.; Han, M.; Xie, B. Strain sensors based on chromium nanoparticle arrays. *Nanoscale* **2014**, *6*, 3930–3933. [CrossRef] [PubMed]

176. Li, S.; Lin, M.M.; Toprak, M.S.; Kim, K.; Muhammed, M. Nanocomposites of polymer and inorganic nanoparticles for optical and magnetic applications. *Nano Rev.* **2010**, *1*, 5214. [CrossRef]

177. Kim, J.; Van Der Bruggen, B. The use of nanoparticles in polymeric and ceramic membrane structures: Review of manufacturing procedures and performance improvement for water treatment. *Environ. Pollut.* **2010**, *158*, 2335–2349. [CrossRef]

178. Benetti, G.; Cavaliere, E.; Banfi, F.; Gavioli, L. Antimicrobial nanostructured coatings: A gas phase deposition and magnetron sputtering perspective. *Materials* **2020**, *13*, 784. [CrossRef]

179. Kylián, O.; Popok, V.N. Applications of polymer films with gas-phase aggregated nanoparticles. *Front. Nanosci.* **2020**, *15*, 119–162. [CrossRef]

180. Meyer, R.; Lemire, C.; Shaikhutdinov, S.K.; Freund, H.-J. Surface chemistry of catalysis by gold. *Gold Bull.* **2004**, *37*, 72–124. [CrossRef]

181. Arenz, M.; Gilb, S.; Heiz, U. Size effects in the chemistry of small clusters. *Chem. Phys. Sol. Surf.* **2007**, *12*, 1–51. [CrossRef]

182. Haruta, M. Gold catalysts prepared by coprecipitation for low-temperature oxidation of hydrogen and of carbon monoxide. *J. Catal.* **1989**, *115*, 301–309. [CrossRef]

183. Castleman, A.W.; Bowen, K.H. Clusters: Structure, energetics, and dynamics of intermediate states of matter. *J. Phys. Chem.* **1996**, *100*, 12911–12944. [CrossRef]

184. Heiz, U.; Schneider, W.-D. Nanoassembled model catalysts. *J. Phys. D Appl. Phys.* **2000**, *33*, R85–R102. [CrossRef]

185. Lang, S.; Popolan, D.; Bernhardt, T. Chemical reactivity and catalytic properties of size-selected gas-phase metal clusters. *Chem. Phys. Sol. Surf.* **2007**, *12*, 53–90. [CrossRef]

186. Poppa, H. Nucleation, growth, and TEM analysis of metal particles and clusters deposited in UHV. *Catal. Rev.* **1993**, *35*, 359–398. [CrossRef]

187. Chen, M.; Goodman, D. Oxide-supported metal clusters. *Chem. Phys. Sol. Surf.* **2007**, *12*, 201–269. [CrossRef]

188. Mammen, N.; Spanu, L.; Tyo, E.C.; Yang, B.; Halder, A.; Seifert, S.; Pellin, M.J.; Vajda, S.; Narasimhan, S. Reversing size-dependent trends in the oxidation of copper clusters through support effects. *Eur. J. Inorg. Chem.* **2017**, *2018*, 16–22. [CrossRef]

189. Halder, A.; Vajda, S. Synchrotron characterization of clusters for catalysis. *Front. Nanosci.* **2020**, *15*, 1–29. [CrossRef]

190. Arenz, M.; Landman, U.; Heiz, U. CO combustion on supported gold clusters. *Chem. Phys. Chem.* **2006**, *7*, 1871–1879. [CrossRef]

191. Mayoral, A.; Martínez, L.; Garcia-Martin, J.M.; Fernández-Martinez, I.; Garcia-Hernandez, M.; Galiana, B.; Ballesteros, C.; Huttel, Y. Tuning the size, composition and structure of Au and Co$_{50}$Au$_{50}$ nanoparticles by high-power impulse magnetron sputtering in gas phase synthesis. *Nanotechnology* **2019**, *30*, 065606. [CrossRef]

192. Yang, B.; Liu, C.; Halder, A.; Tyo, E.C.; Martinson, A.B.F.; Seifert, S.; Zapol, P.; Curtiss, L.A.; Vajda, S. Copper cluster size effect in methanol synthesis from CO2. *J. Phys. Chem. C* **2017**, *121*, 10406–10412. [CrossRef]

193. Chen, P.-T.; Tyo, E.C.; Hayashi, M.; Pellin, M.J.; Safonova, O.V.; Nachtegaal, M.; Van Bokhoven, J.A.; Vajda, S.; Zapol, P. Size-selective reactivity of subnanometer Ag4 and Ag16 clusters on a TiO2 surface. *J. Phys. Chem. C* **2017**, *121*, 6614–6625. [CrossRef]

194. Gawande, M.B.; Goswami, A.; Asefa, T.; Guo, H.; Biradar, A.V.; Peng, D.-L.; Zbořil, R.; Varma, R. Core–shell nanoparticles: Synthesis and applications in catalysis and electrocatalysis. *Chem. Soc. Rev.* **2015**, *44*, 7540–7590. [CrossRef] [PubMed]

195. Jang, M.-H.; Kizilkaya, O.; Kropf, A.J.; Kurtz, R.L.; Elam, J.W.; Lei, Y. Synergetic effect on catalytic activity and charge transfer in Pt-Pd bimetallic model catalysts prepared by atomic layer deposition. *J. Chem. Phys.* **2020**, *152*, 024710. [CrossRef] [PubMed]

196. Raja, R.; Hermans, S.; Shephard, D.S.; Johnson, B.F.G.; Sankar, G.; Bromley, S.T.; Thomas, J.M. Preparation and characterisation of a highly active bimetallic (Pd–Ru) nanoparticle heterogeneous catalyst†. *Chem. Commun.* **1999**, 1571–1572. [CrossRef]

197. Negreiros, F.R.; Halder, A.; Yin, C.; Singh, A.; Barcaro, G.; Sementa, L.; Tyo, E.C.; Pellin, M.J.; Bartling, S.; Meiwes-Broer, K.-H.; et al. Bimetallic Ag-Pt sub-nanometer supported clusters as highly efficient and robust oxidation catalysts. *Angew. Chem. Int. Ed.* **2017**, *57*, 1209–1213. [CrossRef]

198. Roy, C.; Sebok, B.; Scott, S.B.; Fiordaliso, E.M.; Sørensen, J.E.; Bodin, A.; Trimarco, D.B.; Damsgaard, C.D.; Vesborg, P.C.K.; Hansen, O.; et al. Impact of nanoparticle size and lattice oxygen on water oxidation on NiFeOxHy. *Nat. Catal.* **2018**, *1*, 820–829. [CrossRef]

199. Bodin, A.; Christoffersen, A.-L.N.; Elkjær, C.F.; Brorson, M.; Kibsgaard, J.; Helveg, S.; Chorkendorff, I. Engineering Ni–Mo–S nanoparticles for hydrodesulfurization. *Nano Lett.* **2018**, *18*, 3454–3460. [CrossRef]

200. Guirado-López, R.A.; Dorantes-Dávila, J.; Pastor, G.M. Orbital magnetism in transition-metal clusters: From hund's rules to bulk quenching. *Phys. Rev. Lett.* **2003**, *90*, 226402. [CrossRef]

201. Peredkov, S.; Neeb, M.; Eberhardt, W.; Meyer, J.; Tombers, M.; Kampschulte, H.; Niedner-Schatteburg, G. Spin and orbital magnetic moments of free nanoparticles. *Phys. Rev. Lett.* **2011**, *107*, 233401. [CrossRef]

202. Cox, A.J.; Louderback, J.G.; Apsel, S.E.; Bloomfield, L.A. Magnetism in 4d-transition metal clusters. *Phys. Rev. B* **1994**, *49*, 12295–12298. [CrossRef] [PubMed]

203. Payne, F.W.; Jiang, W.; Bloomfield, L.A. Magnetism and magnetic isomers in free chromium clusters. *Phys. Rev. Lett.* **2006**, *97*, 193401. [CrossRef] [PubMed]

204. Binns, C.; Blackman, J. Magnetism in isolated clusters. In *Metallic Nanoparticles*; Blackman, J.A., Ed.; Elsevier: Amsterdam, The Netherlands, 2008; pp. 231–275. [CrossRef]

205. Bucher, J.-P.; Khanna, S.N. Magnetism of free and supported metal clusters. In *Quantum Phenomena in Clusters and Nanostructures*; Khanna, S.N., Castleman, A.W., Jr., Eds.; Springer: Berlin, Germany, 2003; pp. 83–137. [CrossRef]

206. Martins, M.; Wurth, W. Magnetic properties of supported metal atoms and clusters. *J. Phys. Condens. Matter* **2016**, *28*, 503002. [CrossRef]

207. Mukherjee, P.; Manchanda, P.; Kumar, P.; Zhou, L.; Kramer, M.J.; Arti, K.; Skomski, R.; Sellmyer, D.; Shield, J.E. Size-induced chemical and magnetic ordering in individual Fe–Au nanoparticles. *ACS Nano* **2014**, *8*, 8113–8120. [CrossRef] [PubMed]

208. Koten, M.A.; Mukherjee, P.; Shield, J.E. Core-shell nanoparticles driven by surface energy differences in the Co-Ag, W-Fe, and Mo-Co systems. *Part. Part. Syst. Charact.* **2015**, *32*, 848–853. [CrossRef]

209. Rui, X.; Shield, J.E.; Sun, Z.; Xu, Y.; Sellmyer, D.J. In-cluster-structured exchange-coupled magnets with high energy densities. *Appl. Phys. Lett.* **2006**, *89*, 122509. [CrossRef]

210. Kharchenko, A.; Lukashevich, M.; Popok, V.N.; Khaibullin, R.; Valeev, V.; Bazarov, V.; Petracic, O.; Wieck, A.D.; Odzhaev, V. Correlation of electronic and magnetic properties of thin polymer layers with cobalt nanoparticles. *Part. Part. Syst. Charact.* **2013**, *30*, 180–184. [CrossRef]

211. Koch, S.A.; Palasantzas, G.; Vystavel, T.; De Hosson, J.T.M.; Binns, C.; Louch, S. Magnetic and structural properties of Co nanocluster thin films. *Phys. Rev. B* **2005**, *71*, 085410. [CrossRef]

212. Zare-Kolsaraki, H.; Hackenbroich, B.; Micklitz, H. Strongly enhanced tunneling magnetoresistance in granular films of Co clusters in a CO 2 matrix: Evidence for a cluster-surface/matrix interaction. *Europhys. Lett.* **2002**, *57*, 866–871. [CrossRef]

213. Binns, C.; Maher, M.J.; Pankhurst, Q.A.; Kechrakos, D.; Trohidou, K.N. Magnetic behavior of nanostructured films assembled from preformed Fe clusters embedded in Ag. *Phys. Rev. B* **2002**, *66*, 184413. [CrossRef]

214. Katz, E.; Willner, I. Integrated nanoparticle-biomolecule hybrid systems: Synthesis, properties, and applications. *Angew. Chem. Int. Ed.* **2004**, *43*, 6042–6108. [CrossRef]

215. Schmelzer, J.W.P.; Brown, S.A.; Wurl, A.; Hyslop, M.; Blaikie, R.J. Finite-size effects in the conductivity of cluster assembled nanostructures. *Phys. Rev. Lett.* **2002**, *88*, 226802. [CrossRef] [PubMed]

216. Lassesson, A.; Schulze, M.; Van Lith, J.; Brown, S.A. Tin oxide nanocluster hydrogen and ammonia sensors. *Nanotechnology* **2007**, *19*, 015502. [CrossRef] [PubMed]

217. Xie, B.; Liu, L.; Peng, X.; Zhang, Y.; Xu, Q.; Zheng, M.; Takiya, T.; Han, M. Optimizing hydrogen sensing behavior by controlling the coverage in Pd nanoparticle films. *J. Phys. Chem. C* **2011**, *115*, 16161–16166. [CrossRef]

218. Barborini, E.; Vinati, S.; Leccardi, M.; Repetto, P.; Bertolini, G.; Rorato, O.; Lorenzelli, L.; DeCarli, M.; Guarnieri, V.; Ducati, C.; et al. Batch fabrication of metal oxide sensors on micro-hotplates. *J. Micromech. Microeng.* **2008**, *18*, 55015. [CrossRef]

219. Chen, M.; Luo, W.; Xu, Z.; Zhang, X.; Xie, B.; Wang, G.; Han, M. An ultrahigh resolution pressure sensor based on percolative metal nanoparticle arrays. *Nat. Commun.* **2019**, *10*, 4024–4029. [CrossRef]

220. Langer, J.; Novikov, S.M.; Liz-Marzán, L.M. Sensing using plasmonic nanostructures and nanoparticles. *Nanotechnology* **2015**, *26*, 322001. [CrossRef]

221. Gong, Y.; Zhou, Y.; He, L.; Xie, B.; Song, F.; Han, M.; Wang, G. Systemically tuning the surface plasmon resonance of high-density silver nanoparticle films. *Eur. Phys. J. D* **2013**, *67*, 87. [CrossRef]

222. Novikov, S.M.; Popok, V.N.; Evlyukhin, A.B.; Muhammad, H.; Morgen, P.; Fiutowski, J.; Beermann, J.; Rubahn, H.-G.; Bozhevolnyi, S.I. Highly stable monocrystalline silver clusters for plasmonic applications. *Langmuir* **2017**, *33*, 6062–6070. [CrossRef]

223. Popok, V.N.; Novikov, S.M.; Lebedinskij, Y.Y.; Markeev, A.M.; Andreev, A.A.; Trunkin, I.N.; Arsenin, A.V.; Volkov, V.S. Gas-aggregated copper nanoparticles with long-term plasmon resonance stability. *Plasmonics* **2020**, in press. [CrossRef]

224. Polonskyi, O.; Kylián, O.; Drabik, M.; Kousal, J.; Solar, P.; Artemenko, A.; Čechvala, J.; Choukourov, A.; Slavinska, D.; Biederman, H. Deposition of Al nanoparticles and their nanocomposites using a gas aggregation cluster source. *J. Mater. Sci.* **2014**, *49*, 3352–3360. [CrossRef]

225. Minnai, C.; Milani, P. Metal-polymer nanocomposite with stable plasmonic tuning under cyclic strain conditions. *Appl. Phys. Lett.* **2015**, *107*, 73106. [CrossRef]

226. Karas, M.; Bachmann, D.; Hillenkamp, F. Influence of the wavelength in high-irradiance ultraviolet laser desorption mass spectrometry of organic molecules. *Anal. Chem.* **1985**, *57*, 2935–2939. [CrossRef]

227. Tanaka, K.; Waki, H.; Ido, Y.; Akita, S.; Yoshida, Y.; Yoshida, T.; Matsuo, T. Protein and polymer analyses up to m/z 100 000 by laser ionization time-of-flight mass spectrometry. *Rapid Commun. Mass Spectrom.* **1988**, *2*, 151–153. [CrossRef]

228. Hillenkamp, F.; Karas, M.; Beavis, R.C.; Chait, B.T. Matrix-assisted laser desorption/ionization mass spectrometry of biopolymers. *Anal. Chem.* **1991**, *63*, 1193A–1203A. [CrossRef] [PubMed]

229. Karas, M.; Krüger, R. Ion formation in MALDI: The cluster ionization mechanism. *Chem. Rev.* **2003**, *103*, 427–440. [CrossRef] [PubMed]

230. Dreisewerd, K. Recent methodological advances in MALDI mass spectrometry. *Anal. Bioanal. Chem.* **2014**, *406*, 2261–2278. [CrossRef] [PubMed]

231. Chen, C.-H. Mass spectrometry for forensic applications. In *Encyclopedia of Analytical Chemistry*; John Wiley & Sons: Chichester, UK, 2000; pp. 1–15. [CrossRef]

232. Brodbelt, J.S.; Hildenbrand, Z.L.; Schug, K.A. Applications of MALDI-TOF MS in environmental microbiology. *Analyst* **2016**, *141*, 2827–2837. [CrossRef]

233. Duncan, M.W.; Nedelkov, D.; Walsh, R.; Hattan, S.J. Applications of MALDI mass spectrometry in clinical chemistry. *Clin. Chem.* **2016**, *62*, 134–143. [CrossRef]

234. Seng, P.; Rolain, J.-M.; Fournier, P.-E.; La Scola, B.; Drancourt, M.; Raoult, D. MALDI-TOF-mass spectrometry applications in clinical microbiology. *Future Microbiol.* **2010**, *5*, 1733–1754. [CrossRef]

235. Croxatto, A.; Prod'Hom, G.; Greub, G. Applications of MALDI-TOF mass spectrometry in clinical diagnostic microbiology. *FEMS Microbiol. Rev.* **2012**, *36*, 380–407. [CrossRef] [PubMed]

236. Pavlovic, M.; Huber, I.; Konrad, R.; Busch, U. Application of MALDI-TOF MS for the identification of food borne bacteria. *Open Microbiol. J.* **2013**, *7*, 135–141. [CrossRef] [PubMed]

237. Parker, C.E.; Pearson, T.W.; Anderson, N.L.; Borchers, C.H. Mass-spectrometry-based clinical proteomics—A review and prospective. *Analyst* **2010**, *135*, 1830–1838. [CrossRef] [PubMed]

238. Cleland, T.P.; Schroeter, E.R. A comparison of common mass spectrometry approaches for paleoproteomics. *J. Proteome Res.* **2018**, *17*, 936–945. [CrossRef]

239. Castellana, E.T.; Sherrod, S.D.; Russell, D.H. Nanoparticles for selective laser desorption/ionization in mass spectrometry. *J. Lab. Autom.* **2008**, *13*, 330–334. [CrossRef]

240. Yonezawa, T.; Kawasaki, H.; Tarui, A.; Watanabe, T.; Arakawa, R.; Shimada, T.; Mafuné, F. Detailed investigation on the possibility of nanoparticles of various metal elements for surface-assisted laser desorption/ionization mass spectrometry. *Anal. Sci.* **2009**, *25*, 339–346. [CrossRef]

241. Hayasaka, T.; Goto-Inoue, N.; Zaima, N.; Shrivas, K.; Kashiwagi, Y.; Yamamoto, M.; Nakamoto, M.; Setou, M. Imaging mass spectrometry with silver naeoparticles reveals the distribution of fatty acids in mouse retinal sections. *J. Am. Soc. Mass Spectrom.* **2010**, *21*, 1446–1454. [CrossRef]

242. Abdelhamid, H.N. Nanoparticle assisted laser desorption/ionization mass spectrometry for small molecule analytes. *Microchim. Acta* **2018**, *185*, 200. [CrossRef]

243. Ng, K.-M.; Chau, S.-L.; Tang, H.-W.; Wei, X.-G.; Lau, K.-C.; Ye, F.; Ng, A.M.C. Ion-desorption efficiency and internal-energy transfer in surface-assisted laser desorption/ionization: More implication(s) for the thermal-driven and phase-transition-driven desorption process. *J. Phys. Chem. C* **2015**, *119*, 23708–23720. [CrossRef]

244. Li, Y.; Cao, X.; Zhan, L.; Xue, J.; Wang, J.; Xiong, C.; Nie, Z. Hot electron transfer promotes ion production in plasmonic metal nanostructure assisted laser desorption ionization mass spectrometry. *Chem. Commun.* **2018**, *54*, 10905–10908. [CrossRef] [PubMed]

245. Prysiazhnyi, V.; Dycka, F.; Kratochvíl, J.; Stranak, V.; Popok, V.N. Effect of Ag nanoparticle size on ion formation in nanoparticle assisted LDI MS. *Appl. Nano* **2020**, *1*, 2. [CrossRef]

246. Prysiazhnyi, V.; Dycka, F.; Kratochvíl, J.; Stranak, V.; Ksirova, P.; Hubicka, Z. Silver nanoparticles for solvent-free detection of small molecules and mass-to-charge calibration of laser desorption/ionization mass spectrometry. *J. Vac. Sci. Technol. B* **2019**, *37*, 012906. [CrossRef]

247. Prysiazhnyi, V.; Dycka, F.; Kratochvil, J.; Sterba, J.; Stranak, V. Gas-aggregated Ag nanoparticles for detection of small molecules using LDI MS. *Anal. Bioanal. Chem.* **2019**, *412*, 1037–1047. [CrossRef] [PubMed]

248. Muller, L.; Kailas, A.; Jackson, S.N.; Roux, A.; Barbacci, D.C.; Schultz, J.A.; Balaban, C.D.; Woods, A.S. Lipid imaging within the normal rat kidney using silver nanoparticles by matrix-assisted laser desorption/ionization mass spectrometry. *Kidney Int.* **2015**, *88*, 186–192. [CrossRef] [PubMed]

249. Jackson, S.N.; Baldwin, K.; Muller, L.; Womack, V.M.; Schultz, J.A.; Balaban, C.D.; Woods, A.S. Imaging of lipids in rat heart by MALDI-MS with silver nanoparticles. *Anal. Bioanal. Chem.* **2013**, *406*, 1377–1386. [CrossRef] [PubMed]

250. Muller, L.; Baldwin, K.; Barbacci, D.C.; Jackson, S.N.; Roux, A.; Balaban, C.D.; Brinson, B.E.; McCully, M.I.; Lewis, E.K.; Schultz, J.A.; et al. Laser desorption/ionization mass spectrometric imaging of endogenous lipids from rat brain tissue implanted with silver nanoparticles. *J. Am. Soc. Mass Spectrom.* **2017**, *28*, 1716–1728. [CrossRef]

251. Corbelli, G.; Ghisleri, C.; Marelli, M.; Milani, P.; Ravagnan, L. Highly deformable nanostructured elastomeric electrodes with improving conductivity upon cyclical stretching. *Adv. Mater.* **2011**, *23*, 4504–4508. [CrossRef]

252. Gnatkovsky, V.; Cattalini, A.; Antonini, A.; Spreafico, L.; Saini, M.; Noe, F.; Alessi, C.; Librizzi, L.; Uva, L.; Marras, C.E.; et al. Recording electrical brain activity with novel stretchable electrodes based on supersonic cluster beam implantation nanotechnology on conformable polymers. *Int. J. Nanomed.* **2019**, *14*, 10079–10089. [CrossRef]

253. Popok, V.N. Ion implantation of polymers: Formation of nanoparticulate materials. *Rev. Adv. Mater. Sci.* **2012**, *30*, 1–26.

254. Minnai, C.; Di Vece, M.; Milani, P. Mechanical-optical-electro modulation by stretching a polymer-metal nanocomposite. *Nanotechnology* **2017**, *28*, 355702. [CrossRef]

255. Potenza, M.A.C.; Nazzari, D.; Cremonesi, L.; Denti, I.; Milani, P. Hyperspectral imaging with deformable gratings fabricated with metal-elastomer nanocomposites. *Rev. Sci. Instrum.* **2017**, *88*, 113105. [CrossRef] [PubMed]

256. Aqra, F.; Ayyad, A. Surface free energy of alkali and transition metal nanoparticles. *Appl. Surf. Sci.* **2014**, *314*, 308–313. [CrossRef]

257. Ruffino, F.; Torrisi, V.; Marletta, G.; Grimaldi, M.G. Effects of the embedding kinetics on the surface nano-morphology of nano-grained Au and Ag films on PS and PMMA layers annealed above the glass transition temperature. *Appl. Phys. A* **2012**, *107*, 669–683. [CrossRef]

258. Bonde, H.C.; Fojan, P.; Popok, V.N. Controllable embedding of size-selected copper nanoparticles into polymer films. *Plasma Process. Polym.* **2020**, *17*, 1900237. [CrossRef]

259. Hanif, M.; Juluri, R.R.; Fojan, P.; Popok, V.N. Polymer films with size-selected silver nanoparticles as plasmon resonance-based transducers for protein sensing. *Biointerface Res. Appl. Chem.* **2016**, *6*, 1564–1568.

260. Popok, V.N.; Muhammad, H.; Ceynowa, F.; Fojan, P. Immersion of low-energy deposited metal clusters into poly(methyl methacrylate). *Nucl. Instrum. Methods Phys. Res. B* **2017**, *409*, 91–95. [CrossRef]

261. Ceynowa, F.A.; Chirumamilla, M.; Zenin, V.A.; Popok, V.N. Arrays of size-selected metal nanoparticles formed by cluster ion beam technique. *MRS Adv.* **2018**, *3*, 2771–2776. [CrossRef]

262. Kuzminova, A.; Beranová, J.; Polonskyi, O.; Shelemin, A.; Kylián, O.; Choukourov, A.; Slavinska, D.; Biederman, H. Antibacterial nanocomposite coatings produced by means of gas aggregation source of silver nanoparticles. *Surf. Coat. Technol.* **2016**, *294*, 225–230. [CrossRef]

263. Kylián, O.; Kratochvíl, J.; Petr, M.; Kuzminova, A.; Slavinska, D.; Biederman, H.; Beranová, J. Ag/C:F antibacterial and hydrophobic nanocomposite coatings. *Funct. Mater. Lett.* **2017**, *10*, 1750029. [CrossRef]

264. Kratochvíl, J.; Sterba, J.; Lieskovská, J.; Langhansová, H.; Kuzminova, A.; Khalakhan, I.; Kylián, O.; Stranak, V. Antibacterial effect of Cu/C:F nanocomposites deposited on PEEK substrates. *Mater. Lett.* **2018**, *230*, 96–99. [CrossRef]

265. Kratochvíl, J.; Kuzminova, A.; Kylián, O. State-of-the-art, and perspectives of, silver/plasma polymer antibacterial nanocomposites. *Antibiotics* **2018**, *7*, 78. [CrossRef] [PubMed]

266. Kylián, O.; Polonskyi, O.; Kratochvíl, J.; Artemenko, A.; Choukourov, A.; Drabik, M.; Solar, P.; Slavinska, D.; Biederman, H.; Choukourov, A. Control of wettability of plasma polymers by application of Ti nano-clusters. *Plasma Process. Polym.* **2011**, *9*, 180–187. [CrossRef]

267. Kuzminova, A.; Shelemin, A.; Kylián, O.; Petr, M.; Kratochvíl, J.; Solar, P.; Biederman, H. From super-hydrophilic to super-hydrophobic surfaces using plasma polymerization combined with gas aggregation source of nanoparticles. *Vacuum* **2014**, *110*, 58–61. [CrossRef]

268. Kylián, O.; Kuzminova, A.; Hanuš, J.; Slavinska, D.; Biederman, H. Super-hydrophilic SiOx coatings prepared by plasma enhanced chemical vapor deposition combined with gas aggregation source of nanoparticles. *Mater. Lett.* **2018**, *227*, 5–8. [CrossRef]

269. Shelemin, A.; Nikitin, D.; Choukourov, A.; Kylián, O.; Kousal, J.; Khalakhan, I.; Melnichuk, I.; Slavinska, D.; Biederman, H.; Choukourov, A. Preparation of biomimetic nano-structured films with multi-scale roughness. *J. Phys. D Appl. Phys.* **2016**, *49*, 254001. [CrossRef]

270. Choukourov, A.; Kylián, O.; Petr, M.; Vaidulych, M.; Nikitin, D.; Hanuš, J.; Artemenko, A.; Shelemin, A.; Gordeev, I.; Kolská, Z.; et al. RMS roughness-independent tuning of surface wettability by tailoring silver nanoparticles with a fluorocarbon plasma polymer. *Nanoscale* **2017**, *9*, 2616–2625. [CrossRef]

271. Kylián, O.; Petr, M.; Serov, A.; Solar, P.; Polonskyi, O.; Hanuš, J.; Choukourov, A.; Biederman, H. Hydrophobic and super-hydrophobic coatings based on nanoparticles overcoated by fluorocarbon plasma polymer. *Vacuum* **2014**, *100*, 57–60. [CrossRef]

272. Petr, M.; Hanuš, J.; Kylián, O.; Kratochvíl, J.; Solar, P.; Slavinska, D.; Biederman, H. Superhydrophobic fluorine-free hierarchical coatings produced by vacuum based method. *Mater. Lett.* **2016**, *167*, 30–33. [CrossRef]

273. Kratochvíl, J.; Kuzminova, A.; Solař, P.; Hanuš, J.; Kylián, O.; Biederman, H. Wetting and drying on gradient-nanostructured C:F surfaces synthesized using a gas aggregation source of nanoparticles combined with magnetron sputtering of polytetrafluoroethylene. *Vacuum* **2019**, *166*, 50–56. [CrossRef]

274. Vaidulych, M.; Shelemin, A.; Hanuš, J.; Khalakhan, I.; Krakovsky, I.; Kočová, P.; Mašková, H.; Kratochvíl, J.; Pleskunov, P.; Štěrba, J.; et al. Superwettable antibacterial textiles for versatile oil/water separation. *Plasma Process. Polym.* **2019**, *16*, 1900003. [CrossRef]

275. Solar, P.; Kylián, O.; Marek, A.; Vandrovcova, M.; Bačáková, L.; Hanuš, J.; Vyskocil, J.; Slavinska, D.; Biederman, H. Particles induced surface nanoroughness of titanium surface and its influence on adhesion of osteoblast-like MG-63 cells. *Appl. Surf. Sci.* **2015**, *324*, 99–105. [CrossRef]

276. Kylián, O.; Kratochvíl, J.; Hanuš, J.; Polonskyi, O.; Solař, P.; Biederman, H. Fabrication of Cu nanoclusters and their use for production of Cu/plasma polymer nanocomposite thin films. *Thin Solid Films* **2014**, *550*, 46–52. [CrossRef]

277. Vahl, A.; Carstens, N.; Strunskus, T.; Faupel, F.; Hassanien, A. Diffusive memristive switching on the nanoscale, from individual nanoparticles towards scalable nanocomposite devices. *Sci. Rep.* **2019**, *9*, 17367. [CrossRef] [PubMed]

278. Polonskyi, O.; Solar, P.; Kylián, O.; Drabik, M.; Artemenko, A.; Kousal, J.; Hanuš, J.; Pešička, J.; Matolínová, I.; Kolíbalová, E.; et al. Nanocomposite metal/plasma polymer films prepared by means of gas aggregation cluster source. *Thin Solid Films* **2012**, *520*, 4155–4162. [CrossRef]

279. Maicu, M.; Schmittgens, R.; Hecker, D.; Glöß, D.; Frach, P.; Gerlach, G. Synthesis and deposition of metal nanoparticles by gas condensation process. *J. Vac. Sci. Technol. A* **2014**, *32*, 2. [CrossRef]

280. Hanuš, J.; Steinhartová, T.; Kylián, O.; Kousal, J.; Malinský, P.; Choukourov, A.; Macková, A.; Biederman, H. Deposition of Cu/a-C:H nanocomposite films. *Plasma Process. Polym.* **2016**, *13*, 879–887. [CrossRef]

281. Vaidulych, M.; Hanuš, J.; Steinhartová, T.; Kylián, O.; Choukourov, A.; Beranová, J.; Khalakhan, I.; Biederman, H. Deposition of Ag/a-C:H nanocomposite films with Ag surface enrichment. *Plasma Process. Polym.* **2017**, *14*, 1600256. [CrossRef]

282. Thukkaram, M.; Vaidulych, M.; Kylián, O.; Hanus, J.; Rigole, P.; Aliakbarshirazi, S.; Asadian, M.; Nikiforov, A.; Van Tongel, A.; Biederman, H.; et al. Investigation of Ag/a-C:H nanocomposite coatings on titanium for orthopedic applications. *ACS Appl. Mater. Interfaces* **2020**, *12*, 23655–23666. [CrossRef]

283. Zeng, H.; Du, X.; Singh, S.C.; Kulinich, S.A.; Yang, S.; He, J.; Cai, W. Nanomaterials via laser ablation/irradiation in liquid: A review. *Adv. Funct. Mater.* **2012**, *22*, 1333–1353. [CrossRef]

284. Streubel, R.; Barcikowski, S.; Gökce, B. Continuous multigram nanoparticle synthesis by high-power, high-repetition-rate ultrafast laser ablation in liquids. *Opt. Lett.* **2016**, *41*, 1486–1489. [CrossRef]

285. Sportelli, M.C.; Izzi, M.; Volpe, A.; Clemente, M.; Picca, R.A.; Ancona, A.; Lugarà, P.M.; Palazzo, G.; Cioffi, N.; Izzi, M. The pros and cons of the use of laser ablation synthesis for the production of silver nano-antimicrobials. *Antibiotics* **2018**, *7*, 67. [CrossRef] [PubMed]

286. Premkumar, P.A.; Starostin, S.A.; De Vries, H.; Paffen, R.M.J.; Creatore, M.; Eijkemans, T.J.; Koenraad, P.M.; Van De Sanden, M. High quality SiO2-like layers by large area atmospheric pressure plasma enhanced CVD: Deposition process studies by surface analysis. *Plasma Process. Polym.* **2009**, *6*, 693–702. [CrossRef]
287. Ding, Y.; Yamada, R.; Gresback, R.; Zhou, S.; Pi, X.; Nozaki, T. A parametric study of non-thermal plasma synthesis of silicon nanoparticles from a chlorinated precursor. *J. Phys. D Appl. Phys.* **2014**, *47*, 485202. [CrossRef]
288. Bilik, N.; Greenberg, B.L.; Yang, J.; Aydil, E.S.; Kortshagen, U.R. Atmospheric-pressure glow plasma synthesis of plasmonic and photoluminescent zinc oxide nanocrystals. *J. Appl. Phys.* **2016**, *119*, 243302. [CrossRef]
289. Mariotti, D.; Sankaran, R.M. Microplasmas for nanomaterials synthesis. *J. Phys. D Appl. Phys.* **2010**, *43*, 323001. [CrossRef]
290. Yu, W.; Xie, H. A review on nanofluids: Preparation, stability mechanisms, and applications. *J. Nanomater.* **2012**, *2012*, 435873. [CrossRef]
291. Taylor, R.A.; Coulombe, S.; Otanicar, T.; Phelan, P.; Gunawan, A.; Lv, W.; Rosengarten, G.; Prasher, R.; Tyagi, H. Small particles, big impacts: A review of the diverse applications of nanofluids. *J. Appl. Phys.* **2013**, *113*, 011301. [CrossRef]

Review

Nanoparticles Synthesised in the Gas-Phase and Their Applications in Sensors: A Review

Evangelos Skotadis *, Evangelos Aslanidis, Maria Kainourgiaki and Dimitris Tsoukalas *

Department of Applied Physics, National Technical University of Athens, 15780 Athens, Greece;
evaaslani@central.ntua.gr (E.A.); mkainourgiaki@mail.ntua.gr (M.K.)
* Correspondence: evskotad@central.ntua.gr (E.S.); dtsouk@central.ntua.gr (D.T.); Tel.: +30-21-077-22-929 (E.S.)

Received: 1 October 2020; Accepted: 27 October 2020; Published: 3 November 2020

Abstract: This article aims to provide a comprehensive review of recent advances in the use of gas-phase synthesized nanoparticles in the field of sensing technology. Since there are numerous and diverse reviews that already cover the subject extensively, this review focuses predominantly but not exclusively on gas-phase synthesized metallic nanoparticles and their most prominent sensing-applications. After a brief overview on the main uses of nanoparticles in science and technology, as well as a description of the dominant fabrication methods, the review discusses their incorporation in strain-sensing, chemical sensing and bio-sensing as well as a few other sensing-applications. The review highlights the inherent advantages of nanoparticles, as well as how they combine with flexible gas-phase synthesis processes.

Keywords: biosensors; chemical sensors; gas phase; gas sensors; nanoparticles; sputtering; strain sensors

---

1. Introduction

Nanotechnology has been one of the main driving forces over the recent past in scientific fields as diverse as material science, physics, electronics, biotechnology, medicine, etc. The main focus of nanotechnology is the implementation of materials with dimensions in the nanometer scale in devices. Macroscopic materials' fundamental properties drastically change when they are shrunk to a size that is comparable to that of DNA, viruses and atoms, highlighting an enormous potential for unique and novel technologies. Nanomaterials can be considered as artificial atoms that can be combined so as to obtain new materials with unprecedented properties in an always expanding field of possible applications.

Organic, inorganic or metallic nanoparticles (NPs), remain to date amongst the most studied of nanomaterials. There is an abundance of original scientific reports as well as reviews related to the properties and possible uses of individual organic NPs, inorganic NPs, NP ensembles, metal oxides, dendrimers, proteins, micelles etc.: targeted drug delivery [1], molecular biotechnology [2], the use of metal oxide nanoparticles as antibacterial agents [3], their combination with polymers [4], their use as nano-sized power generators [5] and in environmental remediation [6], are only a few of their possible applications.

Metallic nanoparticles in particular, are widely and commonly employed as active materials in various applications such as optical sensors [7], in light-based technologies [8], as potent antimicrobial materials [9], in chemical and biological sensing [10], in catalysis [11], in drug delivery [12], in imaging applications [13], in sustainable energy technology [14] and in energy-harvesting [15]. Apart from metallic NPs consisting of a single metal, bimetallic NPs synthesized as alloys or core-shell structures are also used in a wide range of applications [16]. The integration of NPs in sensing technology is of particular interest since their high surface/volume ratio and unique optical, chemical and electrical properties render them as ideal active materials in almost any sensing scenario. Solid Au NPs have been studied extensively as chemiresistive [17] or plasmonic surface enhancement for Raman scattering-based devices [18], for the chemical sensing of volatile organic compounds (VOCs), biosensors [19], sensors for infectious diseases [20], environmental sensing [21], food quality inspection [22] as well as strain sensing [23].

The current review aims to report on recent research efforts on sensing devices, based on NPs grown by physical methods with emphasis on gas-phase grown NPs. Strictly speaking, the gas phase manufacturing technique of bare metallic NPs, utilizes processes that will be described in Section 2 and are illustrated in Figure 1. However, in this review we also consider sensors that make use of bare NPs formed by the two-step process, namely physical deposition and annealing of a thin metal layer. We believe that the latter investigations are of interest for researchers using NP deposition by the gas phase technique since performing the NP fabrication at room temperature (R.T.) and in a single step, can be an attractive alternative method for replacing the two-step process. The article begins by briefly discussing the gas-phase technique for NP synthesis as well as the chemical and biosynthesis of NPs. The second part of the review showcases recent results related to strain-sensors as well as a few other types of devices. In the second part of the review a brief description regarding charge-transport and the strain-sensing mechanism in the case of gas-phase NPs will be also provided; this discussion relates directly to other types of sensors discussed in the following sections of the article. As a next step the review discusses one of the most reported type of sensing-devices employing gas-phase NPs, i.e., chemical or gas-sensors and will finally report on recent progress in the fields of bio-sensing. It is the authors hope that after a thorough review on recent advances in this field, the reader will appreciate the flexibility and the advantages offered by gas-phase generated NPs, in the design of a sensing device.

2. Nanoparticle Synthesis

2.1. Gas-Phase Nanoparticle Synthesis

One of the most common methods for NP synthesis is the one employing the physical vapour deposition (PVD) technique of direct current (DC) or radio frequency (RF) sputtering; this technique does not feature any dangerous chemical reagents or complex chemical synthesis steps. In DC sputtering a metallic target is bombarded by ions of an inert gas; in this way, gas-phase atoms from the target's surface are detached and due to a pressure difference between the nanoparticle generation chamber and the deposition chamber (Figure 1), they start to condensate whilst moving towards the deposition chamber. This process allows for facile NP size control as well as NP surface coverage on top of the deposition substrate; in addition, the entire process is conducted in R.T. Metallic nanoparticles can be also formed using sputtering of a metal target when this is followed by thermal treatment of the deposited metallic film that results in material aggregation and thus NP formation. The sputtering technique is common in the microelectronics industry hence suitable for mass production and batch fabrication, as well as in research. Today, there exists a great number of publications related to sputtering's principles, theoretical studies for particle formation as well as modifications on sputtering systems so as to obtain improved properties [24]. Since a detailed discussion related to sputtering is outside the scope of the current review, readers interested on the subject are strongly encouraged to consult the relevant publications.

2.2. Alternative Methods for NP Synthesis

Perhaps the most common method for NP synthesis as well as shape-control is the one based on chemical methods [25] that have been covered extensively in the literature; such methods produce NPs that are commonly called colloidal. In brief, colloidal NCs are synthesized in a liquid solution containing some stabilizing organic molecules, broadly termed as surfactants. As extracted from their growing medium, they are typically made of a crystalline core with the desired chemical composition and a monolayer surface shell of tightly coordinated surfactants that provide them with solubility and, hence, with colloidal stability. In addition, surfactants can tune the overall physiochemical properties of the NPs. In colloidal synthesis, molecular precursors containing the atomic species that will form the NCs are introduced into a flask and allowed to react or decompose at a suitable temperature [26].

More recently, the green-synthesis of biogenic NPs (NPs deriving from biomaterials) has also attracted attention, by providing cost-effective and easy to prepare NPs that employ eco-friendly materials. Amongst the key advantages that this technique offers over chemical synthesis methods, is the biological capacity to catalyze reactions in aqueous media at standard temperature and pressure [27]. The biological synthesis of nanoparticles is a bottom-up approach that involves the use of living prokaryotic/eukaryotic cells, biomolecules extracted from living cells and cell free supernatant or extracted biomolecules; these can act as reducing or capping agents for NP formation. The biogenic NP synthesis is enabled by key enzymes (e.g., oxidoreductases, NADH-dependent reductases etc.) that are present in biomaterials, which can first bind to metallic ions and then reduce them. Today, many reports can be found on the extracellular synthesis of metallic NPs, by using different bacterial strains, fungal species, yeasts and plant or flower extracts as bio-nanofactories for eco-friendly-one-step-rapid synthesis NP formation [28].

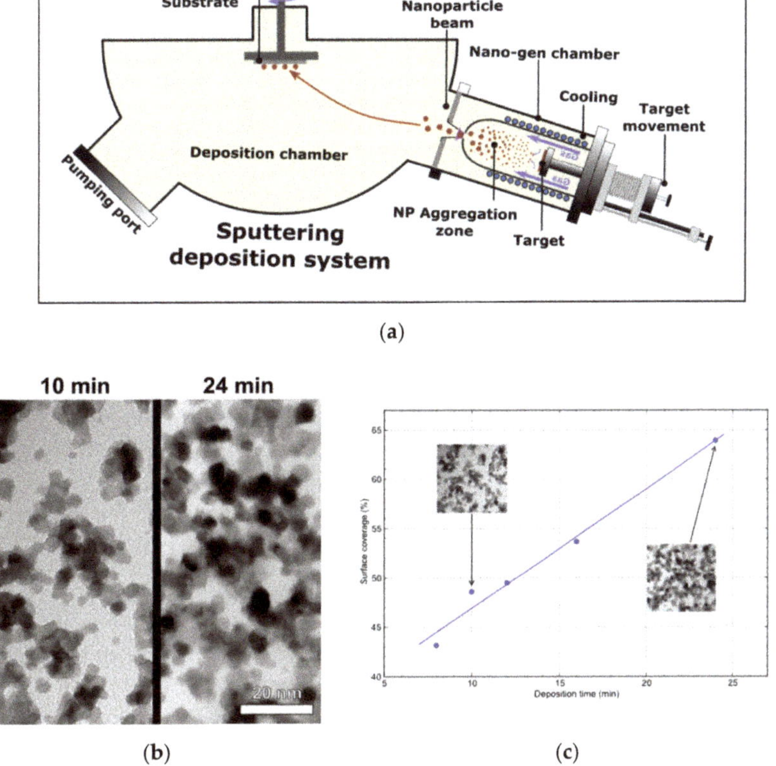

(a)

(b)　(c)

Figure 1. (**a**) Schematic of a typical ion-sputtering system. Key components can be seen in the Deposition and Nano-gen chambers (**b**) Bright field TEM micrographs comparing NP film surface coverage for varying deposition time (10 min and 24 min). (**c**) Surface coverage in relation to deposition time. Figure 1b,c reprinted from [29], with permission from Elsevier.

3. Strain Sensors and Other Devices

This part of the review will deal with recent advances mainly in the field of strain sensing. Since the conductivity of gas-phase, NP-based sensors can be tuned by modifying the surface coverage/density of the NP film on the substrate's surface, devices of varying conductivity can be produced. Strain sensing devices can either utilize NP films with a conductivity that lies just below the percolation threshold (conductivity via an activated tunnelling mechanism), dense "metallic" type films as well as employing cracked substrates (oxides etc.) in order to further enhance the sensitivity (usually indicated by the gauge factor, GF) of the device. All results discussed in the current section are summarized in Table 1. Section 3.1 discusses the physical mechanisms of charge transport phenomena in the case of sensors based on gas-phase synthesized NPs; this discussion is also relevant to devices presented in Sections 4 and 5 of this review.

3.1. Conductivity in Metallic NP Films, Strain-Sensing Mechanism

The operational principle of nanoparticle-based sensors is based on the fact that NPs have an inter-particle distance, which is key to the conductivity of the device. Assuming that the inter-particle distance is defined by l, the conductivity is given by the following equation:

$$R = r_0 \exp(\beta l) \exp(E_c/k_b T) \tag{1}$$

where β is the tunneling constant, r_0 a pre-exponential constant, k_b is the Boltzmann constant, T the temperature and E_c the activation energy which is given by:

$$E_c = (e^2/8\pi\varepsilon\varepsilon_0) \left[1/r - 1/(r + l)\right] \tag{2}$$

where, r is the mean diameter of the nanoparticles and ε the electric permittivity of the dielectric medium. At R.T. E_c (K_b T) $\ll$ 1. Herrmann et al. [30], have proposed a theory suggesting that the differential resistance change is given by the following equation:

$$\Delta R/R = \exp(g\gamma) - 1 \tag{3}$$

where, g is the strain sensitivity or the strain G.F. and γ the deformation. For small γ, Equation (3) becomes:

$$\Delta R/R = g\gamma \tag{4}$$

this is the case for chemically synthesized, solvent-based, cross-linked NP sensors.

Aslanidis et al. [31] have shown that there is a need to modify the previous equations so as to account for changes introduced in the case of gas-phase NP sensors, highlighting that Equation (3) is not adequate to describe their performance. Relative resistance changes in the case of gas-phase NP sensors, display strain ranges that can be described with a linear Equation (4) and others where they are still linear but with a different slope and therefore different sensitivity; this cannot be accounted for by the cross-linked NPs model. In their study, Aslanidis et al. conclude that new inter-particle gaps are formed during device bending. These new gaps are critical for device sensitivity, increasing its G.F. above a specific strain-threshold. In fact, the new G.F. (g) of the device can be calculated by the following equation:

$$\Delta R/R_0 = \gamma \, (g + kr_0/R_0 + k'r_0/R_0) \tag{5}$$

Equation (5) is linear, with different slopes and has proven to fit more accurately the behaviour of gas-phase NP sensors than the previously reported model.

Table 1. Gas-phase NP strain-sensors.

Authors/Year [Ref]	Active Material	Sensor Type	Sensitivity	Notes
Tanner et al./2012 [32]	Pt NPs	Strain-sensor	75 G.F.	Gas-phase NPs, Si substrate
Zheng et al./2014 [33]	Cr NPs	Strain-sensor	100 G.F.	Gas-phase NPs, PET substrate
Xie et al./2018 [34]	Pd NPs	Strain-sensor	1000 G.F.	Gas-phase NPs, PET substrate
Patsiouras et al./2018 [35]	Pt NPs	Strain-sensor	45 G.F.	Gas-phase NPs, Si substrate
Schwebke et al./2018 [36]	Pt NPs	Strain-sensor	23 (BN), 9–9500 (Al_2O_3) G.Fs	Gas-phase NPs, BN & Al_2O_3 substrate
Min et al./2019 [37]	Ag NPs/MWCNTs composites	Strain/stretch-sensor	58.7 G.F.	Gas-phase NPs, PDMS substrate
Lee et al./2019 [38]	Ag NPs	Strain-sensor	290.62, 1056 (cracks)	Gas-phase NPs, PI substrate
Liu et al./2019 [39]	Pd NPs	Strain-sensor	55 (0.3% strain)–3500 (8% strain)	Gas-phase NPs, PET substrate
Aslanidis et al. [31]	Pt NPs	Strain-sensor	60 G.F. (strain < 0.64%), 85 G.F. (strain > 0.64%)	Gas-phase NPs, PI substrate
Chen et al./2019 [40]	Pd NPs	Pressure-sensor/Barometer	0.13 kPa^{-1}	Gas-phase NPs, PET substrate
Zhang et al./2017 [41]	Au NPs/UCNPs structure	Temperature-sensor	1.35% K^{-1} (325 K)	Gas-phase NPs, PI substrate

3.2. Review of Recent Advances in NP-Based, Strain-Sensors and Other Physical Sensors

One of the first published results related to gas-phase NPs was that of Tanner et al. [32]; in that paper the authors took advantage of sputtering's potential for tuning the surface coverage of the device, optimizing the strain sensors performance accordingly. Highest sensitivities have been recorded for intermediate NP densities, just below the percolation threshold, where the tunnelling/hopping charge-transport mechanism is dominant. Zheng et al. [33] have deposited Cr NPs on top of flexible PET substrates. The authors produced devices dominated by quantum charge-transport phenomena, resulting in improved GF compared to semiconducting or metal foil strain sensors, as well as in a wider dynamic range with a workable maximum applied strain beyond 3%. Xie et al. [34] used Pd NPs on top of a PET substrate in three different NP surface densities that were dominated by the hopping/tunnelling mechanism. Again, sensors with lower surface densities exhibited higher sensitivities (G.F. of 1000) as well as mechanical robustness in repeated fatigue tests. Patsiouras et al. [35] utilized thin alumina (Al_2O_3) coatings deposited via the atomic layer deposition technique and on top of Pt NP networks, in order to study the potential of the alumina layer to act as an effective humidity barrier coating. Parameters such as alumina thickness and deposition temperature were evaluated; in addition, the effect of alumina and NP size on the GF was also studied. Schwebke et al. [36] deposited Pt NP films on boron nitride (BN) and alumina films that were previously deposited on polyimide substrates. The application of high strains (close to 1.5%) reveals the formation of cracks in the alumina functionalized strain sensors, which translates in a gigantic resistance jump and a proportional increase in GF (from 9 to 9500). In contrast to the brittle alumina films, softer BN substrates remain linear throughout the applied strain range. Min et al. [37] used an aerodynamically focused nanomaterials (AFN) printing system (a dry printing method) in order to transfer aerosolised NPs/MWCNTs composites onto a PDMS substrate. The authors studied critical fabrication steps such as the printing process and the effect of substrate, achieving high levels of deformation (74%) and good mechanical stability (1000 load/unload strain cycles). Lee et al. [38] studied the AFN fabrication extensively and optimized the AFN printing technique (printing patterns & printing velocity) for the production of Ag NP strain-sensors on polyimide (PI) substrates; this resulted in sensors with a GF between 18.6 and 290.62. In addition, the authors induced mechanical cracks within the Ag NP film, thus radically increasing the GF of the device. Liu et al. [39] studied the effect of Pd NP concentration on sensor sensitivity for sensors implemented on PET substrates. Various GFs have been achieved according to the applied strain range. Finally, Aslanidis et al. [31] reported on alumina coatings used as a protection against humidity, for Pt NP-based flexible sensors (Figure 2). The authors established a critical alumina thickness that is suitable for protection for increased strains (flexible substrates) while at the same time submitting the device in intensive fatigue tests; in addition, they also examined the effect of alumina on sensor sensitivity. Another contribution was the proposition of a model explaining the behaviour of the sensor, as strain gradually increases (introduction of two distinctive GFs); this model highlights the difference between sensors utilizing solvent-free NPs (gas-phase) and solvent-based (colloidal) NPs.

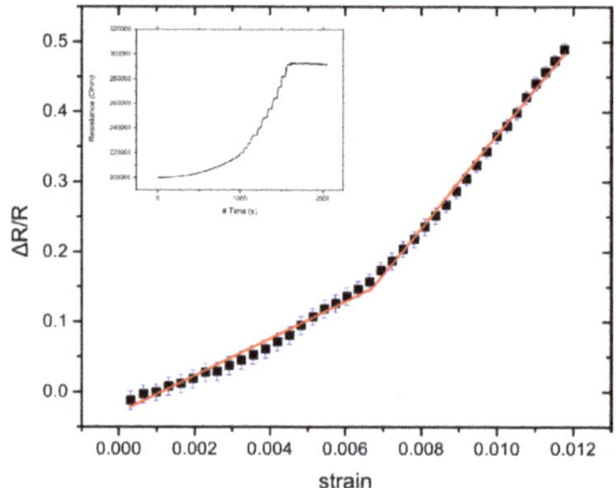

Figure 2. Measurement of resistance during gradual application of strain. Relative resistance over strain graph, where the GF is calculated by the slopes of the fitting line. Error bars represent the standard deviation of the measurements. In this particular example, the sensor has GF1 ~26 for lower strains and GF2 ~66 for higher strains. Inset picture shows the sensor's actual "step-like" response to gradual strain application. Reprinted from [30]

Gas-phase, NP-based sensors are also used for other types of physical sensors. Chen et al. [40] realised a highly sensitive piezoresistive sensor that can also be used as a barometer, by depositing Pd NPs on PET substrates. Contrary to traditional pressure sensors, this device transduces the external pressure on the elastic membrane on which the NPs are deposited; in this way the tunnelling conductance between the percolated NPs changes, resulting in high sensitivity. Zhang et al. [41] deposited Lanthanide-doped (β-NaYF$_4$: Yb^{3+}/Er^{3+}) upconversion nanoparticles (UCNPs), on top of gas-phase Au NPs previously deposited on microfibers. The plasma-resonance on the surface of the Au nanofilm can be excited by launching a 980 nm-wavelength laser beam into the microfiber, resulting in an enhancement of the local electric field and a strong thermal effect. The fabricated device has proved to be able to operate as a temperature sensor.

4. Chemical Sensors

The field of chemical or gas sensing is the one featuring the highest number of scientific publications related to gas-phase metallic NPs. The main body of work related to gas-sensors, deals with conductometric devices that utilize noble metallic nanoparticle as catalysts combined with metal oxides, while other publications rely on the production of a hybrid NP/polymer-based device. The advantage of fabricating and depositing NPs at room temperature (RT) and in a highly controlled manner is a unique feature of gas-phase techniques that have been widely employed in order to produce a vast number of chemical sensors. All results discussed in the current section are summarized in Table 2.

Table 2. Gas-phase NP chemical sensors.

Authors/Year [Ref]	Active Material	Gas Target	LoD (Concentration)	Notes
Lee et al./2016 [42]	Au decorated CuO NWs	CO, NO_2	1 ppm	Heat-treated NPs
Kim et al. [43]	Pt decorated SnO_2 NFs	Toluene	10 ppm	Heat-treated NPs
Wongrat et al./2016 [44]	Au decorated ZnO	EtOH	100 ppm	Heat-treated NPs
Choi et al./2017 [45]	Pt decorated SWCNTs	NO_2	2 ppm	Heat-treated NPs
Yang et al./2017 [46]	Pd decorated SnO_2 on $LiNbO_3$	H_2	100 ppm	Heat-treated NPs
Gasparotto et al./2018 [47]	Au decorated ZnO NRs	H_2, O_2	1996 ppm	Heat-treated NPs
Liang et al./2018 [48]	Au decorated VO_2 NWs	NO_2	5 ppm	Heat-treated NPs
Drmosh et al./2018 [49]	Au decorated SnO_2	NO_2	50 ppm	Heat-treated NPs
Cao et al./2019 [50]	Pd decorated ZnO NRs	EtOH	100 ppm	Heat-treated NPs
Khalid et al./2019 [51]	Au decorated SnO_2 NWs	EtOH	125 ppm	Heat-treated NPs
Jaiswal et al./2020 [52]	Pd decorated MoS_2	H_2	10 ppm	Heat-treated NPs
Liang et al./2016 [53]	Au decorated VO_2 nanosheets	CH_4	100 ppm	Gas-phase NPs
Yuan et al./2016 [54]	Au decorated PS/WO_3 composites	NO_2	50 ppb	Gas-phase NPs
Li et al./2017 [55]	Au & SnO_2 decorated MoS_2	TEA	2 ppm	Gas-phase NPs
Hao et al./2017 [56]	Pd decorated $MoS_2/SiO_2/Si$ heterojunction	H_2	0.5%	Gas-phase NPs
Vernieres et al./2017 [57]	Fe nanocubes	NO_2	3 ppb	Gas-phase NPs
Dhall et al./2017 [58]	Pt decorated $MWCNTs/TiO_2$	H_2	0.05%	Gas-phase NPs
Song et al./2017 [59]	Au decorated ZnO NRs	TEA	1 ppm	Gas-phase NPs
Chen et al./2017 [60]	Pd coated PMMA membranes	H_2	50 ppm	Gas-phase NPs
Arachchige/2018 [61]	Au decorated MoO_3 NFs	H_2S, acetone, EtOH, H_2	ppb level, 5, 0.2, 20 ppm (respectively)	Gas-phase NPs
Xie et al./2018 [62]	Pd NPs on PET	H_2	15 ppm	Gas-phase NPs
Koo et al./2019 [63]	Pt decorated Al-doped ZnO	Acetone	0.1 ppm	Gas-phase NPs
Chen et al./2020 [64]	Au decorated, 3D MoS_2	TEA	2 ppm	Gas-phase NPs
Sysoev et al./2009 [65]	SnO_2 NPs/NWs	2-Propanol	1 ppm	Gas-phase NPs
Shaalan et al./2011 [66]	SnO_2 NPs/ NWs/MWs	NO_2	2 ppm	Gas-phase NPs
Bhatnagar et al./2017 [67]	SnO_2/ SnO_2:C NPs	H_2, EtOH	2%	Gas-phase NPs
Vasiliev et al./2018 [68]	SnO_2 NPs	H_2	20 ppm	Gas-phase NPs
Skotadis et al. /2012 [29]	Pt NPs with single polymer coating	EtOH, R.H.	500 ppm	Gas-phase NPs
Skotadis et al. /2013 [71]	Pt NPs with single polymer coating	EtOH, RH	500 ppm	Gas-phase NPs
Madianos et al./2018 [72]	Pt NPs with four polymer coatings	Chlorpyrifos, RH.	100 ppb (chlorpyrifos)	Gas-phase NPs
Skotadis et al. /2020 [73]	Pt NPs with four polymer coatings	Chloract 48 EC, RH	73.95 ppb (chlorpyrifos)	Gas-phase NPs
Afify et al./2015 [74]	W^{4+} NPs/sepiolite grains	RH	40%	Precipitated NPs
Hassan et al./2016 [75]	ZnO NPs/sepiolite needles	RH, NO_2 and H_2	28%, ppm levels, 20 ppm	Precipitated NPs

In an effort to improve sensitivity and selectivity of multiply-networked p-type CuO nanowires (NWs), Lee et al. [42] functionalized NWs with heat-treated Au NPs of varying size. Heat treating thin, RF sputtered, metallic films, in a range usually between 400 and 500 °C, is one of the most common methods for obtaining well-dispersed and uniform metallic NPs. By mainly changing the thickness of the initial metal layer, the NP size can be tuned. The heat-treatment approach has been followed by many other groups so as to functionalize metal oxides with NPs and in order to improve their, sensitivity, selectivity and reduce their operating temperature (from 150–400 °C) to RT. It is worth noting that in most cases this two-step process that requires heating of the device at high temperatures, could be replaced by the single-step, RT gas-phase synthesis method. In any case, the use of high surface to volume materials such as NPs increases the active surface of the device by introducing additional catalysis sites. To be more specific, Kim et al. [43] have demonstrated improved sensitivity for toluene by modifying SnO_2 nanofibres (NFs) using Pt NP layers of varying thickness; Wongrat et al. [44] employed Au NPs to improve the sensitivity of ZnO nanostructures towards ethanol detection; Choi et al. [45] modified single walled carbon nanotubes (SWCNTs) with Pt NPs for better sensitivity, recovery time and selectivity; Yang et al. [46] achieved better hydrogen detection by modifying SnO_2 thin films on top of a $LiNbO_3$ piezoelectric substrate with Pd NPs; Gasparotto et al. [47] improved the sensitivity of ZnO nanorods (NRs) towards H_2 and O_2 detection by decorating them with Au NPs; Liang et al. [48] employed Au-decorated VO_2 NWs of varying Au size and surface concentration to study their effect on sensitivity for NO_2 detection; Drmosh et al. [49] deposited Au NPs on SnO_2 films for improved sensitivity, in RT NO_2 detection; Cao et al. [50] used Pd NPs decorated ZnO NRs, for improved ethanol detection at 260 °C, due to the catalytic behaviour of the Pd layer; Khalid et al. [51] decorated SnO_2 NWs grown on quartz and silicon substrates with Au NPs for improved ethanol detection in RT. Finally, Jaiswal et al. [52] modified MoS_2 with Pd NPs, for improved H_2 sensitivity compared to pristine MoS_2.

Reports on single-step, gas-phase synthesized NPs used in tandem with metal oxides are also abundant; the use of metal oxides, CNTs and 2D nanomaterials in tandem with catalytic noble metallic NPs, guarantees improved sensitivity, selectivity and operating temperature: Liang et al. [53] decorated VO_2 nanosheets with Au NPs, achieving increased sensitivity for methane compared to untreated nanosheets; Yuan et al. [54] have functionalized tungsten oxide/porous silicon (PS) composites with Au NPs, achieving ppb levels of sensitivity in RT; Li et al. [55] deposited Au and SnO_2 NPs on the surface of MoS_2 for triethylamine (TEA) detection, achieving better sensor properties compared to pristine MoS_2; Hao et al. [56] decorated few-layered $MoS_2/SiO_2/Si$ heterojunctions with Pd nanoparticles, increasing its sensitivity towards H_2 by optimizing Pd thickness; Vernieres et al. [57] fabricated Fe Nanocubes for NO_2 detection, this yielded ppb level sensitivity by operating the device on a micro-hotplate; Dhall et al. [58] introduced Pt NPs in a $MWCNTs/TiO_2$ hybrid material, thus achieving improved sensitivity towards H_2 in RT; Song et al. [59] decorated ZnO nanorods grown directly on flat Al_2O_3 ceramic electrodes with Au NPs, for the successful detection of TEA in close to RT conditions (40 °C) and for a relative humidity (RH) of 30%. Chen et al. [60] coated PMMA membranes with Pd NPs to fabricate a H_2 sensor that has a tunable sensitivity; the sensor can be optimized by tuning the thickness of the PMMA layer and NP size. Arachchige et al. [61] decorated molybdenum trioxide (MoO_3) NFs with Au NPs towards the detection of H_2S, acetone, EtOH and H_2 gases; the sensor showed increased sensitivity (10 times higher) if compared to pristine MoO_3. Xie et al. [62] fabricated an optically transparent and flexible H_2 sensor based on Pd NPs on top of a polyethylene terephthalate (PET) sheet; the sensor was evaluated under bending and was able to operate in a wide range of R.H. Koo et al. [63] reported on Pt-decorated Al-doped ZnO (Pt-AZO) NPs for optimized acetone detection; the Pt catalytic effect leads to an increase of absorbed oxygen ion species on the material's surface. Chen et al. [64] deposited Au nanoparticles on the surface of 3D MoS_2 nanostructures, previously grown on top of Al_2O_3 tubular nanostructures, thus fabricating a TEA gas sensor.

Bare (without any decoration) metal-oxide NPs have been also used as chemical sensors. Sysoev et al. [65] used pristine SnO_2 NPs as well as SnO_2 nanowires (NWs), for the detection of 2-propanol; the NW networks have been found to show better stability as well as sensitivity. Shaalan et al. [66] fabricated SnO_2 NP films, micro-wires (MWs) as well NWs, in order to compare their response towards NO_2 vapours. In all cases the operating temperature range of the device was from RT to 300 °C, with peak sensitivity at 150 V for the NWs sensor; the NW sensor outperformed other sensor types due its reduced inter-grain boundaries density. Bhatnagar et al. [67] used pristine SnO_2 NPs as well as SnO_2:C NP alloys, to improve the sensor's selectivity towards H_2 and EtOH (both reducing gases) detection; the sensors operated in a wide temperature range namely: RT −210 °C. Vasiliev et al. [68] used the spark-discharge method to prepare SnO_2 NPs; this "dry" NP synthesis method reduces the hydroxyl concentration on the NP surface, rendering the sensor lees sensitive to any changes in ambient humidity.

In a different approach, other research groups aim to tune the inter-NP distance of the NP layer while the sensor is exposed to volatile organic compounds (VOCs). This transduction process takes advantage of sputtering's ability to produce 2-D, gas-phase, self-assembled NP films of varying surface-coverage, in a repeatable and controllable manner. As a result of this process, devices that lie just below the percolation threshold present conductivity that is governed from quantum charge-transport mechanisms such as tunneling, hopping etc. This renders the devices extremely sensitive in any change in the inter-NP distance. Based on their original experience regarding the production of fully ink-jet printed chemical sensors [69,70], Skotadis et al. [29] combined gas-phase, DC sputtered Pt NP films, with polymer coatings in order to produce gas-sensitive devices (Figure 3). The polymer coatings act as the gas sensitive layer of the device, absorbing many of the VOCs surrounding the sensor. This translates in an increase of device resistance since the swelling of the polymer layer leads to the deformation of the underlying NP film and in an increase in inter-NP distance; the sensing mechanism for this type of devices can be explained by the discussion found in Section 3.1. The Pt NP films were of varying surface coverage, to optimize the sensor according to the NPs distribution. Additionally, the authors have demonstrated the successful implementation of this sensor on top of a flexible polyimide substrate [71]. In an effort to detect and identify pesticide vapours, Madianos et al. [72] improved the previously reported device architecture by incorporating three additional and distinctive polymer coatings. This improves the selectivity of the sensing array and in combination with statistical analysis methods such as principal component analysis (PCA), the device ultimately identified between pesticide vapours (chlorpyrifos) and RH. In a 2020 publication, Skotadis et al. [73] have used a similar device architecture to identify between a commercially available pesticide, namely Chloract 48 EC (a chlorpyrifos-based pesticide), and RH (Figure 4). Finally, in two publications by Afify et al. [74] and Hassan et al. [75], non-traditional materials such as sepiolite grains and needles decorated by NPs, have been employed as chemical sensors. In [74], various oxide/hydroxide NPs have precipitated on top of sepiolite grains; pellets of the respective material combined with tungsten NPs exhibited the best behaviour as humidity sensors, operating at RT. In [75] ZnO NPs have precipitated on top of sepiolite needles, towards their implementation as humidity, NO_2 and H_2 chemical sensors, operating at an optimal temperature of 300 °C.

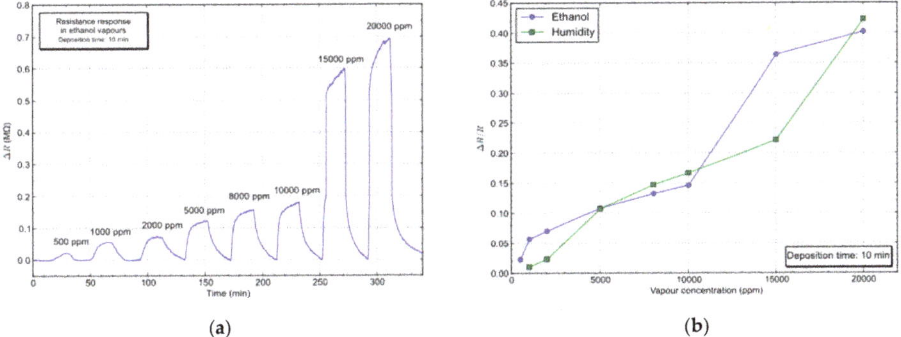

(a)   (b)

Figure 3. (**a**) Transient resistance response to humidity, for a hybrid Pt NP/polymer chemical-sensor, realized after a nanoparticle deposition of 10 min in ethanol (**b**). Relative resistance response of the same sensor in humidity and ethanol. Reprinted from [29], with permission from Elsevier.

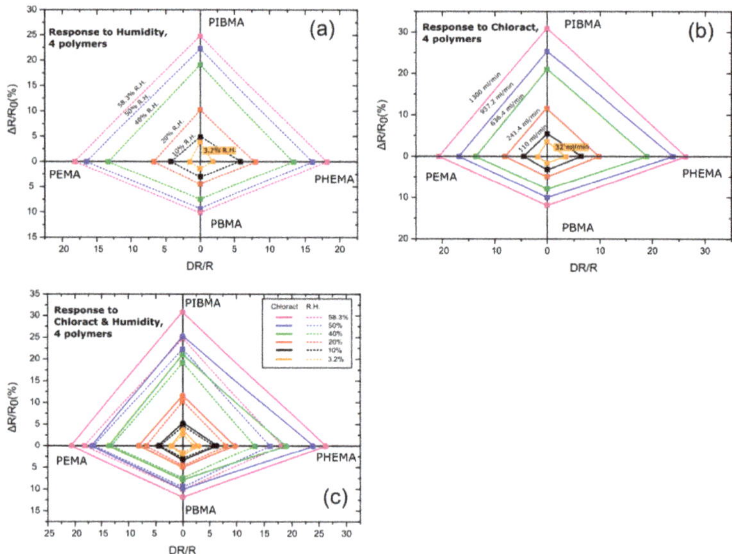

Figure 4. Distinctive fingerprint-patterns after the response of hybrid Pt NPs/polymer based sensors with all four different polymeric coatings towards (**a**) relative humidity (RH) from 3.2% to 58.3% (**b**) Chloract test-solution, for N_2 carrier-gas flows from 32 to 1300 mL/min (identical N_2 flows with the ones used to achieve R.H. from 3.2% to 58.3%) and (**c**) combined response to both analytes. Reprinted from [73], with permission from Elsevier.

5. Biosensing Devices

In the final part of this review, biosensing devices based on gas-phase NPs will be discussed. As in the case of chemical sensors, varying materials such as metal oxides, 2D materials etc. have been also combined with catalytic NPs in order to produce biosensing devices. High aspect ratio and catalytic properties of noble gas-phase NPs are again prominent; it is worth noting that in the case of biosensors, NPs prove to be an exceptional anchoring site for the immobilization of critical materials. All results discussed in the current section are summarized in Table 3.

Hou et al. [76] have reported on ZnO nanorods decorated with Au NPs (ZnO NRs-Au NPs) hybrids, without the addition of any surfactant or reducing agent, for the electrochemical detection of ascorbic acid and uric acid (cancer biomarkers); the Influence of Au NP size and density has been also studied. Li et al. [77] have prepared a disposable electrochemical sensor for the determination of bisphenol A (an endocrine disruptor); the sensor was fabricated by depositing Au NPs on paper and then modifying them with MWCNTs. Skotadis et al. [78] prepared a network of 2D, self-assembled Pt NPs for the detection of DNA hybridization events (even for single base-pair mismatch); hybridized DNA acts as a molecular conductive-bridging between NPs, enhancing conductivity (Figure 5). By modifying the inter-particle distance (controlled via the sputtering deposition time) device sensitivity was optimized. Yuan et al. [79] modified chemical vapour deposition (CVD)-grown graphene (GN) with gas-phase Pt NPs on top of a glassy carbon electrode and without the need for GN transfer; the well-exposed active PT catalyst sites, showed advantages in H_2O_2 detection. In a second article by Yuan et al. [80], the development of CVD grown monolayer GN (on top of glassy carbon) decorated with Au NPs was discussed. The paper reports on the amperometric and non-enzymatic determination of H_2O_2 and avoids any NP contamination due to the transfer-free fabrication process. In a 2017 article, Skotadis et al. [81] prepared an environmental sensor for Pb^{2+} detection in water; the sensor utilizes DNAzymes' disassociation in the presence of Pb^{2+} ions, resulting in a drop of conductivity. Galdino et al. [82] prepared a colorimetric biosensor for total cholesterol detection by using Ionic liquid

(IL)-functionalized graphene oxide (GO), combined with cholesterol oxidase and decorated with Au NPs. The use of gas-phase NPs prevents the undesired by-products and remaining ligands that are always present after chemical NP synthesis. Gasparotto et al. [83] used ZnO NRs decorated by Au NPs, for the electrochemical detection of the ovarian cancer antigen CA-125/MUC126. The Au NPs provide a favourable platform for efficient loading of anti-CA-125 antibody via binding with cystamine and glutaraldehyde (observed by SEM images); the reported limit of detection (LOD) is 100 times lower than a standard immunoblot system. Madianos et al. [84,85] reported on two different configurations for pesticide detection, by immobilizing appropriate aptamers for pesticide capturing on top of Pt NPs. At first, the authors fabricated Pt NP microwires by employing e-beam lithography and DC sputtering; in a follow-up publication the authors fabricated 2D NP networks (less complicated fabrication approach). In both cases target capturing leads in a shift in the sensor's impedance. Biasotto et al. [86] used ZnO NRs modified with gas-phase Au NPs, for efficient Hepatitis C Virus detection. Au was used for critical anti-HCV antibodies-capturing (by coating the Au/ZnO structure with cystamine and glutaraldehyde). Danielson et al. [87] used Au NPs as anchoring sites for biomolecules (streptavidin and DNA) in ZnO NWs. In 2020 Danielson et al. [88] modified GN surfaces using gas-phase Au NPs so as to facilitate the attachment and binding of a DNA aptamer. The sensor operates as a FET and is capable of detecting DNA hybridization events as well as the protein streptavidin. Della Ventura et al. [89] have used the laser-ablation technique (using fs laser pulses) for Au NPs preparation, in order to enhance the sensitivity of a quartz crystal microbalance electrode. The authors have optimized the sensor based on NP surface coverage by successfully detecting antibody-antigen interactions, increasing by 4 times the response of the sensor if compared to a bare QCM electrode.

Table 3. Gas-phase NP biosensors.

Authors/Year [Ref]	Active Material	Target Type	LoD (Concentration)	Notes
Hou et al./2016 [76]	Au decorated ZnO NRs	ascorbic acid, uric acid	0.1 mM, 0.01 mM (respectively)	Gas-phase NPs
Li et al./2016 [77]	Au decorated MWCNTs	bisphenol A	0.03 mg/L	Gas-phase NPs & paper substrate
Skotadis et al./2016 [78]	Pt NPs-2D films	DNA hybridization	1 nM	Gas-phase NPs
Yuan et al./2017 [79]	Pt decorated graphene	H_2O_2	0.18 nM	Gas-phase NPs
Yuan et al./2017 [80]	Au decorated graphene	H_2O_2	10 nM	Gas-phase NPs
Skotadis et al./2017 [81]	Pt NPs-2D films	Pb^{2+} ions	10 nM	Gas-phase NPs
Galdino et al./2017 [82]	Au decorated Graphene Oxide	Total Cholesterol	25 μmol/L	Gas-phase NPs
Gasparotto et al./2017 [83]	Au decorated ZnO NRs	antigen CA-125/MUC126	2.5 ng/μL	Gas-phase NPs
Madianos et al./2018 [84]	Pt NP microwires	Acetamiprid, atrazine	1 pM, 10 pM (respectively)	Gas-phase NPs
Madianos et al./2018 [85]	Pt NPs-2D films	Acetamiprid, atrazine	6 pM, 40 pM (respectively)	Gas-phase NPs
Biasotto et al./2019 [86]	Au decorated ZnO NRs	Hep. C Virus	0.25 μg/μL	Gas-phase NPs
Danielson et al./2019 [87]	Au decorated ZnO NWs	DNA hybridization, streptavidin	100 pM, 10 nM (respectively)	Gas-phase NPs
Danielson et al./2020 [88]	Au decorated graphene	DNA hybridization, streptavidin	15 aM	Gas-phase NPs
Della Ventura et al. [89]	Au NPs	IgG antigen	25 μg/mL	Laser-ablated NPs
Said et al./2017 [90]	Ag/Cu decorated graphene FETs	Glucose	1 μM	Gas-phase NPs
Jung et al./2018 [91]	NiO decorated ZnO NRs FETs	Glucose	0.001 mM	Gas-phase NPs
Soganci et.al./2018 [92]	Cu decorated graphene	Glucose	0.01 mM	Gas-phase NPs
Olejnik et al./2020 [93]	Au decorated Ti	Glucose	30 μM	Heat-treated NPs
Zhang et al. 2020 [94]	Au decorated electrodes	Glucose	1 μM	Gas-phase NPs
Soganci et.al./2020 [95]	Cu NPs/graphene, sandwiched structure	Glucose	0.025 μM	Gas-phase NPs

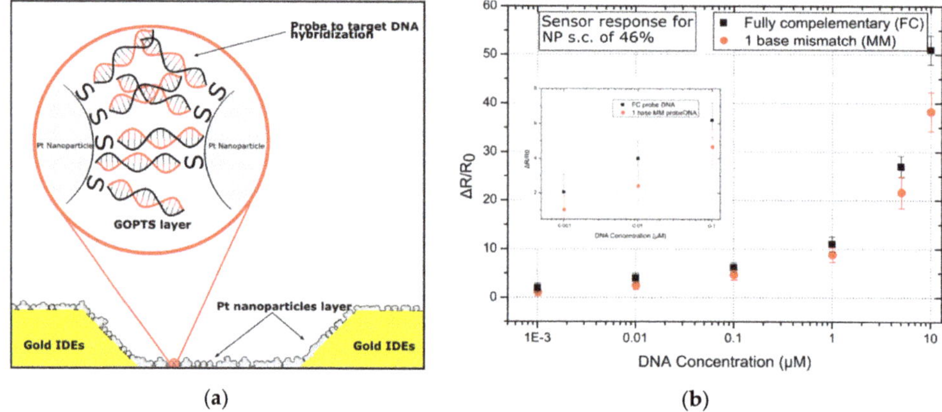

Figure 5. (**a**) Cross section of the Pt NP sensing device (schematic) for DNA hybridization detection, after the target to probe DNA hybridization event (**b**) Sensor response (nanoparticle surface coverage: 46%, IDE gap: 5 μm) for fully complementary (FC) probe to target DNA and single base pair mismatched (1 base MM) probe to target DNA hybridization, in a logarithmic scale. Reprinted from [78], with permission from Elsevier.

It is worth noting that there are many reports focusing on the detection of glucose. If the fact that 20% of the world population suffers from diabetes is taken into account, the importance of fast, selective, portable, low-power, low-cost and accurate glucose sensors is more than obvious. Many efforts focus on the detection of glucose oxide (enzymatic approach) while others focus on its non-enzymatic detection (avoiding expensive and without long-term stability enzymes). Said et al. [90] prepared Graphite-oxide based field-effect transistors (FETs) whose sensitivity was improved by incorporating Ag and Cu NPs on a non-enzymatic sensor. Jung et al. [91] prepared a nonenzymatic flexible FET based glucose sensor using nickel oxide NPs (NiO)-modified zinc oxide nanorods (ZnO NRs). It is worth noting that the sensor was tried in human blood samples measuring glucose efficiently. Soganci et.al. [92] used a GN film prepared by CVD and transferred to FTO glass; the film was then functionalized using Cu NPs, producing a non-enzymatic sensor. Olejnik et al. [93] presented a flexible and enzyme-free electrochemical biosensor based on the combination of Ti and Au NPs, modified by a perm-selective Nafion membrane so as to improve selectivity. Zhang et al. [94] covered two electrode materials (Au and conductive carbon) with 3.4 nm in width Au NPs for the preparation of an enzymatic electrochemical biosensor; the effect of Au NP aerial density was studied by modifying sputtering time. Soganci et.al. [95] in a 2020 article presented a non-enzymatic biosensor based on Cu NPs "sandwiched" between two single GN layers. The unique sandwich structure, presented increased sensitivity if compared to single layer GN decorated NPs; its stability, sensitivity and response time were improved, in accordance with previous theoretical studies, due to the protection of the NP layer by the sandwich structure.

6. Conclusions

A comprehensive review on recent advances of gas-phase synthesized NP-based sensors has been conducted, focusing on strain-sensors, chemical sensors and biological sensors. The flexibility offered by gas-phase synthesis methods offers unique advantages in the development and optimization of nanomaterial-based sensing systems. Simple operation, low-cost, batch fabrication, facile control over many critical device parameters (e.g., NP size and distribution/density), easy material combination for the production of nanoheterostructures, easy integration of biomaterials and their room temperature operation are but a few of their advantages. In addition, the avoidance of any complex chemical processes for the fabrication as well as functionalization of NPs, leaves the surface free of any contaminants and chemical reagent residues; this proves to be critical in many of the aforementioned sensing applications.

If the typical advantages of NPs are additionally taken into account (high surface to volume ratio, catalytic activity, tunable size, tunable device conductivity that often relies on quantum charge transport phenomena, size comparable to that of many biomaterials etc.), the gas-phase synthesis of NPs emerges as a powerful and enabling technology that even years after its initial implementation, still manages to push the boundaries in many and diverse scientific fields. The future prospects of the gas-phase technique remain optimistic since it emerges as the most promising candidate for easy and trouble-free integration in mass-production processes. It is worth noting that especially sputtering has both a significant tradition as well as a strong presence in today's microelectronics industry. Competing techniques such as chemical or especially green synthesis of NPs have yet to establish themselves as suitable and more favourable alternatives for the batch-fabrication of nanoparticle-based devices. Finally, strain-sensors produced via gas-phase synthesis techniques feature enhanced sensitivity when compared to competing techniques. This is partly owed to the technique's flexibility as well as its easy combination with various thin-film deposition methods for the production of extremely sensitive devices. Such results could be easily applied in other sensing applications (e.g., chemical sensors etc.), highlighting the technique's ability to produce advanced and efficient sensing devices.

Author Contributions: E.S. has conducted an overview of recent research papers related to chemical, strain and bio sensors based on Gas-phase synthesized nanoparticles. He has authored/drafted the paper and has also prepared all Figures, Tables etc. E.A. has contributed in finding available papers related to strain sensors. M.K. has contributed in finding available papers related to bio-sensors. D.T. copyedited the paper and has contributed in its structure, scope and purpose as well as in finding specific publications. All authors have read and agreed to the published version of the manuscript.

Funding: This research received no external funding.

Conflicts of Interest: The authors declare no conflict of interest.

References

1. Majidinia, M.; Mirza-Aghazadeh-Attari, M.; Rahimi, M.; Safa, A.; Yousefi, B. Overcoming multidrug resistance in cancer: Recent progress in nanotechnology and new horizons. *IUBMB Life* **2020**, *72*, 855–871. [CrossRef]

2. Niemeyer, C.M. Nanoparticles, proteins, and nucleic acids: Biotechnology meets materials science. *Angew. Chem. Int.* **2001**, *40*, 4128–4158. [CrossRef]

3. Hajipour, M.J.; Fromm, K.M.; Ashkarran, A.A.; De Aberasturi, D.J.; De Larramendi, I.R.; Rojo, T. Antibacterial properties of nanoparticles. *Trends Biotechnol.* **2012**, *30*, 1–13. [CrossRef] [PubMed]

4. Gangopadhyay, R.; De, A. Conducting polymer nanocomposites: A brief overview. *Chem. Mater.* **2000**, *12*, 608–622. [CrossRef]

5. Wang, S.; Lin, L.; Wang, Z.L. Triboelectric nanogenerators as self-powered active sensors. *Nano Energy* **2015**, *11*, 436–462. [CrossRef]

6. Ali, I. New generation adsorbents for water treatment. *Chem. Rev.* **2012**, *112*, 5073–5091. [CrossRef] [PubMed]

7. Anker, J.N.; Hall, W.P.; Lyandres, O.; Shah, N.C.; Zhao, J.; Van Duyne, R.P. Biosensing with plasmonic nanosensors. *Nat. Mater.* **2008**, *7*, 442–453. [CrossRef]

8. Gordon, R. Nanostructured metals for light-based technologies. *Nanotechnology* **2019**, *30*, 212001. [CrossRef]

9. Rai, M.; Yadav, A.; Gade, A. Silver nanoparticles as a new generation of antimicrobials. *Biotechnol. Adv.* **2009**, *27*, 76–83. [CrossRef] [PubMed]

10. Saha, K.; Agasti, S.S.; Kim, C.; Li, X.; Rotello, V.M. Gold nanoparticles in chemical and biological sensing. *Chem. Rev.* **2012**, *112*, 2739–2779. [CrossRef]

11. Astruc, D.; Lu, F.; Aranzaes, J.R. Nanoparticles as recyclable catalysts: The frontier between homogeneous and heterogeneous catalysis. *Angew. Chem. Int.* **2005**, *44*, 7852–7872. [CrossRef]

12. Ghosh, P.; Gang, G.; DE, M.; Kim, C.K.; Rotello, V.M. Gold nanoparticles in delivery applications. *Adv. Drug Deliv. Rev.* **2008**, *60*, 1307–1315. [CrossRef] [PubMed]

13. Sun, C.; Lee, J.S.H.; Zhang, M. Magnetic nanoparticles in MR imaging and drug delivery. *Adv. Drug Deliv. Rev.* **2008**, *60*, 1252–1265. [CrossRef] [PubMed]
14. Maake, P.J.; Bolokang, A.S.; Arendse, C.J.; Iwuoha, E.I.; Motaung, D.E. Metal oxides and noble metals application in organic solar cells. *Sol. Energy* **2020**, *207*, 347–366. [CrossRef]
15. Patnaik, S.; Sahoo, D.P.; Parida, K. An overview on Ag modified g-C3N4 based nanostructured materials for energy and environmental applications. *Renew. Sust. Energ. Rev.* **2018**, *82*, 1297–1312. [CrossRef]
16. Durán, N.; Seabra, A.B. Biogenic synthesized Ag/Au nanoparticles: Production, characterization, and applications. *Curr. Nanosci.* **2018**, *14*, 82–94. [CrossRef]
17. Ibañez, F.J.; Zamborini, F.P. Chemiresistive sensing with chemically modified metal and alloy nanoparticles. *Small* **2012**, *8*, 174–202. [CrossRef] [PubMed]
18. Mahmoud, M.A.; O'Neil, D.; El-Sayed, M.A. Hollow and solid metallic nanoparticles in sensing and in nanocatalysis. *Chem. Mater* **2014**, *26*, 44–58. [CrossRef]
19. Cho, I.-H.; Kim, D.H.; Park, S. Electrochemical biosensors: Perspective on functional nanomaterials for on-site analysis. *Biomater. Res.* **2020**, *24*, 1–12. [CrossRef] [PubMed]
20. Alafeef, M.; Moitra, P.; Pan, D. Nano-enabled sensing approaches for pathogenic bacterial detection. *Biosens. Bioelectron.* **2020**, *165*, 112276. [CrossRef] [PubMed]
21. Liu, B.; Zhuang, J.; Wei, G. Recent advances in the design of colorimetric sensors for environmental monitoring. *Environ. Sci.* **2020**, *7*, 2195–2213. [CrossRef]
22. He, S.; Yuan, Y.; Nag, A.; Mukhopadhyay, S.C.; Organ, D.R. A review on the use of impedimetric sensors for the inspection of food quality. *Int. J. Environ. Res. Public Health* **2020**, *17*, 5220. [CrossRef]
23. Wang, B.; Facchetti, A. Mechanically Flexible Conductors for Stretchable and Wearable E-Skin and E-Textile Devices. *Adv. Mater.* **2019**, *31*, 1901408. [CrossRef]
24. Grammatikopoulos, P.; Steinhauer, S.; Vernieres, J.; Singh, V.; Sowwan, M. Nanoparticle design by gas-phase synthesis. *Adv. Phys. X* **2016**, *1*, 81–100. [CrossRef]
25. Xia, Y.; Xiong, Y.; Lim, B.; Skrabalak, S.E. Shape-controlled synthesis of metal nanocrystals: Simple chemistry meets complex physics? *Angew. Chem. Int.* **2009**, *48*, 60–103. [CrossRef]
26. Cozzoli, P.D.; Manna, L. Synthetic strategies to size and shape controlled nanocrystals and nanocrystal Heterostructures. In *Bio-Applications of Nanoparticles*, 1st ed.; Chan, W.C.W., Ed.; Springer: New York, NY, USA, 2007; Volume 62, pp. 1–14.
27. Schröfel, A.; Kratošová, G.; Šafařík, I.; Raška, I.; Shor, L.M. Applications of biosynthesized metallic nanoparticles—A review. *Acta Biomater.* **2014**, *10*, 4023–4042. [CrossRef]
28. Rana, A.; Yadav, K.; Jagadevan, S. A comprehensive review on green synthesis of nature-inspired metal nanoparticles: Mechanism, application and toxicity. *J. Clean. Prod.* **2020**, *272*, 122880. [CrossRef]
29. Skotadis, E.; Tanner, J.L.; Stathopoulos, S.; Tsouti, V.; Tsoukalas, D. Chemical sensing based on double layer PHEMA polymer and platinum nanoparticle films. *Sens. Actuators B Chem.* **2012**, *175*, 85–91. [CrossRef]
30. Herrmann, J.; Müller, K.H.; Reda, T.; Baxter, G.R.; Raguse, B.D.; De Groot, G.J.; Chai, R.; Roberts, M.; Wieczorek, L. Nanoparticle films as sensitive strain gauges. *Appl. Phys. Lett.* **2007**, *91*, 183105. [CrossRef]
31. Aslanidis, E.; Skotadis, E.; Moutoulas, E.; Tsoukalas, D. Thin film protected flexible nanoparticle strain sensors: Experiments and modelling. *Sensors* **2020**, *20*, 2584. [CrossRef]
32. Tanner, J.L.; Mousadakos, D.; Giannakopoulos, K.; Skotadis, E.; Tsoukalas, D. High strain sensitivity controlled by the surface density of platinum nanoparticles. *Nanotechnology* **2012**, *23*, 285501. [CrossRef]
33. Zheng, M.; Li, W.; Xu, M.; Han, M.; Xie, B. Strain sensors based on chromium nanoparticle arrays. *Nanoscale* **2014**, *6*, 3930–3933. [CrossRef]
34. Xie, B.; Mao, P.; Chen, M.; Liu, J.M.; Wang, G. A tunable palladium nanoparticle film-based strain sensor in a Mott variable-range hopping regime. *Sens. Actuat. A Phys.* **2018**, *272*, 161–169. [CrossRef]
35. Patsiouras, L.; Skotadis, E.; Gialama, N.; Giannakopoulos, K.; Tsoukalas, D. Atomic layer deposited Al2O3 thin films as humidity barrier coatings for nanoparticle-based strain sensors. *Nanotechnology* **2018**, *29*, 465706. [CrossRef]
36. Schwebke, S.; Winter, S.; Koch, M.; Schultes, G. Piezoresistive granular metal thin films of platinum-boron nitride and platinum-alumina at higher strain levels. *J. Appl. Phys.* **2018**, *124*, 235308. [CrossRef]
37. Min, S.H.; Lee, G.Y.; Ahn, S.H. Direct printing of highly sensitive, stretchable, and durable strain sensor based on silver nanoparticles/multi-walled carbon nanotubes composites. *Compos. B. Eng.* **2019**, *161*, 395–401. [CrossRef]

38. Lee, G.Y.; Kim, M.S.; Min, S.H.; Ihn, J.B.; Ahn, S.H. Highly Sensitive Solvent-free Silver Nanoparticle Strain Sensors with Tunable Sensitivity Created Using an Aerodynamically Focused Nanoparticle Printer. *ACS Appl. Mater. Interfaces* **2019**, *11*, 26421–26432. [CrossRef] [PubMed]

39. Liu, F.; Shao, W.; Xu, G.; Yuan, L. Response characteristics of strain sensors based on closely spaced nanocluster films with controlled coverage. *Chin. J. Chem. Phys.* **2019**, *32*, 213–217. [CrossRef]

40. Chen, M.; Luo, W.; Xu, Z.; Wang, G.; Han, M. An ultrahigh resolution pressure sensor based on percolative metal nanoparticle arrays. *Nat. Commun.* **2019**, *10*, 4024. [CrossRef]

41. Zhang, W.; Li, J.; Lei, H.; Li, B. Plasmon-Induced Selective Enhancement of Green Emission in Lanthanide-Doped Nanoparticles. *ACS Appl. Mater. Interfaces* **2017**, *9*, 42935–42942. [CrossRef]

42. Lee, J.S.; Katoch, A.; Kim, J.H.; Kim, S.S. Effect of Au nanoparticle size on the gas-sensing performance of p-CuO nanowires. *Sens. Actuat. B Chem.* **2016**, *222*, 307–314. [CrossRef]

43. Kim, J.H.; Abideen, Z.U.; Zheng, Y.; Kim, S.S. Improvement of toluene-sensing performance of SnO_2 nanofibers by pt functionalization. *Sensors* **2016**, *16*, 1857. [CrossRef]

44. Wongrat, E.; Chanlek, N.; Chueaiarrom, C.; Hongsith, N.; Choopun, S. Low temperature ethanol response enhancement of ZnO nanostructures sensor decorated with gold nanoparticles exposed to UV illumination. *Sens. Actuat. A Phys.* **2016**, *251*, 188–197. [CrossRef]

45. Choi, S.W.; Kim, J.; Byun, Y.T. Highly sensitive and selective NO_2 detection by Pt nanoparticles-decorated single-walled carbon nanotubes and the underlying sensing mechanism. *Sens. Actuat. B Chem.* **2017**, *238*, 1032–1042. [CrossRef]

46. Yang, L.; Yin, C.; Zhang, Z.; Zhou, J.; Xu, H. The investigation of hydrogen gas sensing properties of SAW gas sensor based on palladium surface modified SnO_2 thin film. *Mater. Sci. Semicond. Process.* **2017**, *60*, 16–28. [CrossRef]

47. Gasparotto, G.; Da Silva, R.A.; Zaghete, M.A.; Perazolli, L.A.; Mazon, T. Novel route for fabrication of ZnO nanorods-Au nanoparticles hybrids directly supported on substrate and their application as gas sensors. *Mater. Res.* **2018**, *21*, 20170796. [CrossRef]

48. Liang, J.; Zhu, K.; Yang, R.; Hu, M. Room temperature NO_2 sensing properties of Au-decorated vanadium oxide nanowires sensor. *Ceram. Int.* **2018**, *44*, 2261–2268. [CrossRef]

49. Drmosh, Q.A.; Yamani, Z.H.; Mohamedkhair, A.K.; Hossain, M.K.; Ibrahim, A. Gold nanoparticles incorporated SnO_2 thin film: Highly responsive and selective detection of NO_2 at room temperature. *Mat. Lett.* **2018**, *214*, 283–286. [CrossRef]

50. Cao, P.; Yang, Z.; Navale, S.T.; Stadler, F.J.; Zhu, D. Ethanol sensing behavior of Pd-nanoparticles decorated ZnO-nanorod based chemiresistive gas sensors. *Sens. Actuators B Chem.* **2019**, *298*, 126850. [CrossRef]

51. Wafaa Khalid, K.; Ali Abadi, A.; Abdulqader Dawood, F. Synthesis of SnO_2 Nanowires on Quartz and Silicon Substrates for Gas Sensors. *J. Inorg. Organomet. Polym.* **2020**, *30*, 3294–3304. [CrossRef]

52. Jaiswal, J.; Tiwari, P.; Singh, P.; Chandra, R. Fabrication of highly responsive room temperature H_2 sensor based on vertically aligned edge-oriented MoS_2 nanostructured thin film functionalized by Pd nanoparticles. *Sens. Actuators B Chem.* **2020**, *325*, 128800. [CrossRef]

53. Liang, J.; Li, W.; Liu, J.; Hu, M. Room temperature CH_4 sensing properties of Au decorated VO_2 nanosheets. *Mat. Lett.* **2016**, *184*, 92–95. [CrossRef]

54. Yuan, L.; Hu, M.; Wei, Y.; Ma, W. Enhanced NO_2 sensing characteristics of Au modified porous silicon/thorn-sphere-like tungsten oxide composites. *Appl. Surf. Sci.* **2016**, *389*, 824–834. [CrossRef]

55. Li, W.; Xu, H.; Zhai, T.; Wang, J.; Cao, B. Enhanced triethylamine sensing properties by designing Au@SnO_2/MoS_2 nanostructure directly on alumina tubes. *Sens. Actuators B Chem.* **2017**, *253*, 97–107. [CrossRef]

56. Hao, L.; Liu, Y.; Du, Y.; Xu, Z.; Zhu, J. Highly Enhanced H_2 Sensing Performance of Few-Layer MoS_2/SiO_2/Si Heterojunctions by Surface Decoration of Pd Nanoparticles. *Nanoscale Res. Lett.* **2017**, *12*, 567. [CrossRef]

57. Vernieres, J.; Steinhauer, S.; Zhao, J.; Grammatikopoulos, P.; Sowwan, M. Gas Phase Synthesis of Multifunctional Fe-Based Nanocubes. *Adv. Funct. Mater.* **2017**, *27*, 1605328. [CrossRef]

58. Dhall, S.; Sood, K.; Nathawat, R. Room temperature hydrogen gas sensors of functionalized carbon nanotubes based hybrid nanostructure: Role of Pt sputtered nanoparticles. *Int. J. Hydrogen Energy* **2017**, *42*, 8392–8398. [CrossRef]

59. Song, X.; Xu, Q.; Xu, H.; Cao, B. Highly sensitive gold-decorated zinc oxide nanorods sensor for triethylamine working at near room temperature. *J. Colloid Interf. Sci.* **2017**, *499*, 67–75. [CrossRef] [PubMed]

60. Munasinghe Arachchige, H.M.M.; Zappa, D.; Poli, N.; Gunawardhana, N.; Comini, E. Gold functionalized MoO3 nano flakes for gas sensing applications. *Sens. Actuators B Chem.* **2018**, *269*, 331–339. [CrossRef]

61. Xie, B.; Mao, P.; Chen, M.; Liu, J.M.; Wang, G. Pd Nanoparticle Film on a Polymer Substrate for Transparent and Flexible Hydrogen Sensors. *ACS Appl. Mater. Interfaces* **2018**, *10*, 44603–44613. [CrossRef]

62. Koo, A.; Yoo, R.; Woo, S.P.; Lee, H.S.; Lee, W. Enhanced acetone-sensing properties of pt-decorated al-doped ZnO nanoparticles. *Sens. Actuators B Chem.* **2019**, *280*, 109–119. [CrossRef]

63. Chen, Z.; Xu, H.; Liu, C.; Wang, J.; Cao, B. Good triethylamine sensing properties of Au MoS$_2$ nanostructures directly grown on ceramic tubes. *Mater. Chem. Phys.* **2020**, *245*, 122683. [CrossRef]

64. Chen, M.; Mao, P.; Qin, Y.; Liu, J.M.; Wang, G. Response Characteristics of Hydrogen Sensors Based on PMMA-Membrane-Coated Palladium Nanoparticle Films. *ACS Appl. Mater. Interfaces* **2017**, *9*, 27193–27201. [CrossRef]

65. Sysoev, V.V.; Schneider, T.; Goschnick, J.; Strelcov, E.; Kolmakov, A. Percolating SnO$_2$ nanowire network as a stable gas sensor: Direct comparison of long-term performance versus SnO$_2$ nanoparticle films. *Sens. Actuators B Chem.* **2009**, *139*, 699–703. [CrossRef]

66. Shaalan, N.M.; Yamazaki, T.; Kikuta, T. Influence of morphology and structure geometry on NO$_2$ gas-sensing characteristics of SnO$_2$ nanostructures synthesized via a thermal evaporation method. *Sens. Actuators B Chem.* **2011**, *153*, 11–16. [CrossRef]

67. Bhatnagar, M.; Dhall, S.; Kaushik, V.; Kaushal, A.; Mehta, B.R. Improved selectivity of SnO$_2$: C alloy nanoparticles towards H$_2$ and ethanol reducing gases; role of SnO$_2$: C electronic interaction. *Sens. Actuators B Chem.* **2017**, *246*, 336–343. [CrossRef]

68. Vasiliev, A.A.; Varfolomeev, A.E.; Volkov, I.A.; Jahatspanian, I.E.; Maeder, T. Reducing humidity response of gas sensors for medical applications: Use of spark discharge synthesis of metal oxide nanoparticles. *Sensors* **2018**, *18*, 2600. [CrossRef]

69. Skotadis, E.; Tang, J.; Tsouti, V.; Tsoukalas, D. Chemiresistive sensor fabricated by the sequential ink-jet printing deposition of a gold nanoparticle and polymer layer. *Microelectron. Eng.* **2010**, *87*, 2258–2263. [CrossRef]

70. Tang, J.; Skotadis, E.; Stathopoulos, S.; Roussi, V.; Tsouti, V.; Tsoukalas, D. PHEMA functionalization of gold nanoparticles for vapor sensing: Chemi-resistance, chemi-capacitance and chemi-impedance. *Sens. Actuators B Chem.* **2012**, *2170*, 129–136. [CrossRef]

71. Skotadis, E.; Mousadakos, D.; Katsabrokou, K.; Stathopoulos, S.; Tsoukalas, D. Flexible polyimide chemical sensors using platinum nanoparticles. *Sens. Actuators B Chem.* **2013**, *189*, 106–112. [CrossRef]

72. Madianos, L.; Skotadis, E.; Patsiouras, L.; Filippidou, M.K.; Chatzandroulis, S.; Tsoukalas, D. Nanoparticle based gas-sensing array for pesticide detection. *J. Environ. Chem. Eng.* **2018**, *6*, 6641–6646. [CrossRef]

73. Skotadis, E.; Kanaris, A.; Aslanidis, E.; Michalis, P.; Kalatzis, N.; Chatzipapadopoulos, F.; Marianos, N.; Tsoukalas, D. A sensing approach for automated and real-time pesticide detection in the scope of smart-farming. *Comput. Electron. Agric.* **2020**, *78*, 105759. [CrossRef]

74. Afify, A.S.; Hassan, M.; Piumetti, M.; Peter, I.; Bonelli, B.; Tulliani, J.M. Elaboration and characterization of modified sepiolites and their humidity sensing features for environmental monitoring. *Appl. Clay Sci.* **2015**, *115*, 165–173. [CrossRef]

75. Hassan, M.; Afify, A.S.; Tulliani, J.M. Synthesis of ZnO Nanoparticles onto Sepiolite Needles and Determination of Their Sensitivity toward Humidity. NO$_2$ and H$_2$. *J. Mater. Sci. Technol.* **2016**, *32*, 573–582. [CrossRef]

76. Hou, C.; Liu, H.; Zhang, D.; Yang, C.; Zhang, M. Synthesis of ZnO nanorods-Au nanoparticles hybrids via in-situ plasma sputtering-assisted method for simultaneous electrochemical sensing of ascorbic acid and uric acid. *J. Alloy. Compd.* **2016**, *666*, 178–184. [CrossRef]

77. Li, H.; Wang, W.; Lv, Q.; Bai, H.; Zhang, Q. Disposable paper-based electrochemical sensor based on stacked gold nanoparticles supported carbon nanotubes for the determination of bisphenol. *Electrochem. Commun.* **2016**, *68*, 104–107. [CrossRef]

78. Skotadis, E.; Voutyras, K.; Chatzipetrou, M.; Tsekenis, G.; Patsiouras, L.; Madianos, L.; Chatzandroulis, S.; Zergioti, I.; Tsoukalas, D. Label-free DNA biosensor based on resistance change of platinum nanoparticles assemblies. *Biosens. Bioelectron.* **2016**, *81*, 388–394. [CrossRef]

79. Yuan, Y.; Zhang, F.; Wang, H.; Zheng, Y.; Hou, S. Chemical vapor deposition graphene combined with Pt nanoparticles applied in non-enzymatic sensing of ultralow concentrations of hydrogen peroxide. *RSC Adv.* **2017**, *7*, 30542–30547. [CrossRef]

80. Galdino, N.M.; Brehm, G.S.; Bussamara, R.; Abarca, G.; Scholten, J.D. Sputtering deposition of gold nanoparticles onto graphene oxide functionalized with ionic liquids: Biosensor materials for cholesterol detection. *J. Mater. Chem. B* **2017**, *5*, 9482–9486. [CrossRef]

81. Gasparotto, G.; Costa, J.P.C.; Costa, P.I.; Zaghete, M.A.; Mazon, T. Electrochemical immunosensor based on ZnO nanorods-Au nanoparticles nanohybrids for ovarian cancer antigen CA-125 detection. *Mater. Sci. Eng. C* **2017**, *76*, 1240–1247. [CrossRef] [PubMed]

82. Yuan, Y.; Zheng, Y.; Liu, J.; Wang, H.; Hou, S. Non-enzymatic amperometric hydrogen peroxide sensor using a glassy carbon electrode modified with gold nanoparticles deposited on CVD-grown graphene. *Microchim. Acta* **2017**, *184*, 4723–4729. [CrossRef]

83. Skotadis, E.; Tsekenis, G.; Chatzipetrou, M.; Patsiouras, L.; Madianos, L.; Bousoulas, P.; Zergioti, I.; Tsoukalas, D. Heavy metal ion detection using DNAzyme-modified platinum nanoparticle networks. *Sens. Actuators B Chem.* **2017**, *239*, 962–969. [CrossRef]

84. Madianos, L.; Tsekenis, G.; Skotadis, E.; Patsiouras, L.; Tsoukalas, D. A highly sensitive impedimetric aptasensor for the selective detection of acetamiprid and atrazine based on microwires formed by platinum nanoparticles. *Biosens. Bioelectron.* **2018**, *101*, 268–274. [CrossRef]

85. Madianos, L.; Skotadis, E.; Tsekenis, G.; Patsiouras, L.; Tsigkourakos, M.; Tsoukalas, D. Impedimetric nanoparticle aptasensor for selective and label free pesticide detection. *Microelectron. Eng.* **2018**, *189*, 39–45. [CrossRef]

86. Biasotto, G.; Costa, J.P.C.; Costa, P.I.; Zaghete, M.A. ZnO nanorods-gold nanoparticle-based biosensor for detecting hepatitis C. *Appl. Phys. A* **2019**, *125*, 821. [CrossRef]

87. Danielson, E.; Dhamodharan, V.; Porkovich, A.; Yokobayashi, Y.; Sowwan, M. Gas-Phase Synthesis for Label-Free Biosensors: Zinc-Oxide Nanowires Functionalized with Gold Nanoparticles. *Sci. Rep.* **2019**, *9*, 17370. [CrossRef]

88. Danielson, E.; Sontakke, V.A.; Porkovich, A.J.; Yokobayashi, Y.; Sowwan, M. Graphene based field-effect transistor biosensors functionalized using gas-phase synthesized gold nanoparticles. *Sens. Actuators B Chem.* **2020**, *320*, 128432. [CrossRef]

89. Della Ventura, B.; Funari, R.; Anoop, K.K.; Amoruso, S.; Ausanio, G.; Gesuele, F.; Velotta, R.; Altucci, C. Nano-machining of biosensor electrodes through gold nanoparticles deposition produced by femtosecond laser ablation. *Appl. Phys. B* **2015**, *119*, 497–501. [CrossRef]

90. Said, K.; Ayesh, A.I.; Qamhieh, N.N.; Mahmoud, S.T.; Hisaindee, S. Fabrication and characterization of graphite oxide—nanoparticle composite based field effect transistors for non-enzymatic glucose sensor applications. *J. Alloy. Compd.* **2017**, *694*, 1061–1066. [CrossRef]

91. Jung, D.U.J.; Ahmad, R.; Hahn, Y.B. Nonenzymatic flexible field-effect transistor based glucose sensor fabricated using NiO quantum dots modified ZnO nanorods. *J. Colloid. Interf. Sci.* **2018**, *512*, 21–28. [CrossRef]

92. Soganci, T.; Ayranci, R.; Harputlu, E.; Unlu, C.G.; Ak, M. An effective non-enzymatic biosensor platform based on copper nanoparticles decorated by sputtering on CVD graphene. *Sens. Actuators B Chem.* **2018**, *273*, 1501–1507. [CrossRef]

93. Olejnik, A.; Siuzdak, K.; Karczewski, J.; Grochowska, K. A Flexible Nafion Coated Enzyme-free Glucose Sensor Based on Au-dimpled Ti Structures. *Electroanalysis* **2020**, *32*, 323–332. [CrossRef]

94. Zhang, T.; Ran, J.; Ma, C.; Yang, B. A Universal Approach to Enhance Glucose Biosensor Performance by Building Blocks of Au Nanoparticles. *Adv. Mater. Interfaces* **2020**, *7*, 2000227. [CrossRef]

95. Soganci, T.; Ayranci, R.; Unlu, G.; Acet, M.; Ak, M. Designing sandwich-type single-layer graphene decorated by copper nanoparticles for enhanced sensing properties. *J. Phys. D Appl. Phys* **2020**, *53*, 255105. [CrossRef]

Publisher's Note: MDPI stays neutral with regard to jurisdictional claims in published maps and institutional affiliations.

 *applied nano*

Article

Spontaneous Formation of Core@shell Co@Cr Nanoparticles by Gas Phase Synthesis

Jimena Soler-Morala [1,2], Elizabeth M. Jefremovas [1,3], Lidia Martínez [1,*], Álvaro Mayoral [4,5,6], Elena H. Sánchez [7], Jose A. De Toro [7], Elena Navarro [8,9] and Yves Huttel [1]

[1] Instituto de Ciencia de Materiales de Madrid, Consejo Superior de Investigaciones Científicas (ICMM-CSIC), c/Sor Juana Inés de la Cruz, 3, 28049 Madrid, Spain; jimena.soler@imdea.org (J.S.-M.); elizabma@ucm.es (E.M.J.); huttel@icmm.csic.es (Y.H.)

[2] IMDEA Nanociencia, c/Faraday, 9, 28049 Madrid, Spain

[3] Dpto. Ciencias de la Tierra y Física de la Materia Condensada, Universidad de Cantabria, Avda. de los Castros, 48, 39005 Santander, Spain

[4] Institute of Nanoscience and Materials of Aragon (INMA), Spanish National Research Council (CSIC), University of Zaragoza, 12, Calle de Pedro Cerbuna, 50009 Zaragoza, Spain; amayoral@unizar.es

[5] Laboratorio de Microscopias Avanzadas (LMA), University of Zaragoza, 50015 Zaragoza, Spain

[6] Center for High-Resolution Electron Microscopy (ChEM), School of Physical Science and Technology, ShanghaiTech University, 393 Middle Huaxia Road, Pudong, Shanghai 201210, China

[7] Dpto. de Física Aplicada, Instituto Regional de Investigación Científica Aplicada (IRICA), Universidad de Castilla-La Mancha, 13071 Ciudad Real, Spain; Elena.HSanchez@uclm.es (E.H.S.); JoseAngel.Toro@uclm.es (J.A.D.T.)

[8] Instituto de Magnetismo Aplicado (IMA), UCM-ADIF, 28230 Las Rozas, Spain; enavarro@ucm.es

[9] Departamento de Física de Materiales, Universidad Complutense de Madrid (UCM), 28040 Madrid, Spain

* Correspondence: lidia.martinez@icmm.csic.es

Received: 26 October 2020; Accepted: 12 November 2020; Published: 1 December 2020

Abstract: This work presents the gas phase synthesis of CoCr nanoparticles using a magnetron-based gas aggregation source. The effect of the particle size and Co/Cr ratio on the properties of the nanoparticles is investigated. In particular, we report the synthesis of nanoparticles from two alloy targets, $Co_{90}Cr_{10}$ and $Co_{80}Cr_{20}$. In the first case, we observe a size threshold for the spontaneous formation of a segregated core@shell structure, related to the surface to volume ratio. When this ratio is above one, a shell cannot be properly formed, whereas when this ratio decreases below unity the proportion of Cr atoms is high enough to allow the formation of a shell. In the latter case, the segregation of the Cr atoms towards the surface gives rise to the formation of a shell surrounding the Co core. When the proportion of Cr is increased in the target ($Co_{80}Cr_{20}$), a thicker shell is spontaneously formed for a similar nanoparticle size. The magnetic response was evaluated, and the influence of the structure and composition of the nanoparticles is discussed. An enhancement of the global magnetic anisotropy caused by exchange bias and dipolar interactions, which enables the thermal stability of the studied small particles up to relatively large temperatures, is reported.

Keywords: gas-phase synthesis of nanoparticles; cluster sources; core@shell nanoparticles; CoCr nanoparticles

---

1. Introduction

Materials at the nanoscale have an increasing demand due to the drastic changes in their properties caused by size and surface effects [1]. Some well-known examples of these changes in the material properties include optical [2], catalytic [3], thermal [4], and magnetic properties [5]. The reduced coordination of surface atoms and the high fraction of them in relation to volume yield a variety of surface magnetism effects [6,7]. In this novel magnetism, changes occurring in the surface atoms of nanoparticles of transition metals such as cobalt, iron, or nickel, including the relaxation of lattice parameters and increased density of states at the Fermi level due to their modified electronic structure, lead to enhanced atomic magnetic moments [8] with a higher orbital contribution [9–11]. For the nanoparticle (NP) size range where the number of surface atoms is comparable to that of their volume counterparts, any material in contact with the magnetic material influences the magnetic response, for instance through hybridization with the 3d orbitals of the transition metals [9,12]. In addition, new phenomena characteristic of the nanoscale emerge, such as superparamagnetism [13]; giant magnetoresistance due to spin-dependent scattering [14,15]; or, in bimagnetic systems with components of different anisotropy, exchange coupling [16].

The magnetism of CoCr alloys is due to the ferromagnetism of Co, with a critical superparamagnetic size of 9 nm at room temperature (RT), as Cr is an antiferromagnetic material with a Néel temperature of 311 K [17]. Most of the literature on magnetic studies of the CoCr system focuses on thin alternating Co–Cr layers for applications in magnetic recording or microwave sensing [18–21]. However, to our knowledge, there are no fundamental studies on the magnetism of CoCr NPs, whether alloyed or core@shell-structured.

At the nanoscale, structural and compositional details are of prime importance in assessing fundamental properties, as even small impurities can have a huge effect. In this context, Ion Cluster Sources (ICS), also called Gas Aggregation Sources (GAS), have gained increasing interest in the scientific community, as they allow the fabrication of nanoparticles under extremely clean conditions. In brief, ICS are high vacuum or ultra-high vacuum systems where a given material is atomized (by magnetron sputtering, thermal evaporation, arc discharge, laser ablation, etc.). The resulting atoms are then condensed into nanoparticles that are extracted from the ICS in the form of a NP beam, which is then directed to a surface were the NPs can land with a controlled energy. This gas phase synthesis enables the fabrication of pure nanoparticles of much interest for fundamental studies [22,23]. Here, the synthesis of NPs has been performed using a variant of the Haberland-type gas aggregation source [24]. Specifically, the NPs were grown using a multiple ion cluster source (MICS), which is an adaptation where the single magnetron is replaced by three independent magnetrons [25,26]. This modification has been shown to be very effective for the fabrication of nanoparticles with a tuneable chemical composition and structure [27]. In the current work, however, instead of using two magnetrons for the fabrication of the CoCr NPs, CoCr alloy targets were used with variable amounts of each element, and the versatility of the MICS (in particular regarding the injection of argon through different magnetrons) was exploited in order to study the formation of CoCr nanoparticles with different compositions and structures.

The properties of the grown NPs are discussed in the light of the nanoscale ordering of the constituent elements. In addition, the size of the NPs was adjusted in order to modify the ratio of surface-to-volume atoms and study the structural and magnetic properties of the NPs as a function of such a ratio. In this work, the magnetism of the interacting CoCr nanoparticles was evaluated considering a number of compositional and structural variations. The observed enhancement of the global anisotropy of the NPs is discussed in terms of the exchange bias generated, which increases the thermomagnetic stability, thus "beating the superparamagnetic limit" in small-sized Co NPs [16]. These results will be correlated with the chemical and structural characterization carried out by X-ray photoelectron spectroscopy (XPS) and Scanning High-Resolution Transmission Electron Microscopy (HR-STEM) with Electron Energy Loss Spectroscopy (EELS) for chemical identification.

2. Materials and Methods

Fabrication of the nanoparticles. The nanoparticles (NPs) were grown using a multiple ion cluster source (MICS), with 1" magnetrons, loaded with CoCr targets (Testbourne, Basingstoke, UK, purity 99.95%) with variable amounts of Cr ($Co_{90}Cr_{10}$ and $Co_{80}Cr_{20}$). In the case of the $Co_{90}Cr_{10}$ target, the typical power applied to the magnetron was 14 W, with an Ar flux of 5 or 10 standard cubic centimeters per minute (sccm) in the ignited magnetron and a total Ar flux of 80 sccm. In the case of the $Co_{80}Cr_{20}$ target, the typical power applied to the magnetron was 5 W, with an Ar flux of 5 sccm in the ignited magnetron and a total Ar flux of 80 sccm. The base pressure of the system was below 5×10^{-10} mbar.

The NPs produced with the MICS were deposited in an annex chamber (see Figure S1) either on flat Si(100) polished wafers with a roughness below 1 nm (UniversityWafer Inc., South Boston, MA, USA) or on TEM grids made of ultrathin carbon film on holey carbon support film on copper 400-mesh grids (Ted Pella Inc., Redding, CA, USA). The morphological, structural, chemical, and magnetic characterizations of the NPs were performed ex situ via atomic force microscopy (AFM), scanning transmission electron microscopy (STEM), X-ray photoemission spectroscopy (XPS), and a Superconducting QUantum Interference Device (SQUID). Different samples were grown for each characterization technique, adjusting the deposition time depending on the number of NPs required for each technique.

Atomic Force Microscopy (AFM). An Icon Nanoscope (Veeco) microscope in dynamic mode (soft tapping mode) was used for keeping the excitation frequency constant. Commercial AFM tips from Next-Tip were used. These are silicon tips with a capping of Au NPs with a typical radius of less than 3 nm and a resonance frequency (f_0) and spring constant (k) of $\approx$280 kHz and $\approx$42 N/m, respectively. The analysis of the images was carried out using the WSxM software [28] and homemade macros operated in Igor Pro.

Scanning Transmission Electron Microscopy (STEM) observations were performed using a FEI-TITAN X-FEG transmission electron microscope in the STEM mode, operated at 300 kV. The images were acquired using a high-angle annular dark field (HAADF) detector. The microscope was equipped with a monochromator, the Gatan Energy Filter Tridiem 866 ERS; a spherical aberration corrector (CEOS) for the electron probe (C_s-corrected), allowing an effective 0.08 nm spatial resolution; and an energy dispersive X-ray detector for the EDS analysis. The chemical identification of the elements was carried out by Electron Energy Loss Spectroscopy (EELS).

X-ray photoelectron spectroscopy (XPS) was carried out ex situ in a separate UHV chamber, with a base pressure of 10^{-10} mbar, equipped with a hemispherical electron energy analyzer (SPECS Phoibos 150 spectrometer) and a delay-line detector using a monochromatic Al-K_α (1486.74 eV) X-ray source. Spectra were obtained at a 60° emission take-off angle. Wide scans were recorded using an energy step of 0.5 eV and a pass-energy of 40 eV. Higher energy resolution core level spectra were recorded using an energy step of 0.1 eV and a pass-energy of 20 eV at selected core levels: Co 2p, Cr 2p, and O 1s. The absolute binding energies of the photoelectron spectra were determined by referencing the Cr 2p peak at 576.4 ev [29]. Data processing was performed using the CasaXPS software (Casa software Ltd., Cheshire, UK). Ion sputtering was performed by Ar^+ sputtering at 1.4 keV at different sputtering times (0, 15, 75, 135, 255, and 375 min), with a current measured of 0.67 µA.

Magnetic measurements. The magnetic characterization was carried out using a MPMS-5S and a MPMS-XL5 SQUID magnetometer with the EverCool system (Quantum Design), which can provide a maximum field of 50 kOe and with a broad temperature working range from 2 K to 400 K. Hysteresis loops at different temperatures and thermal dependence magnetization after field-cooled (FC, applying a magnetic field while cooling) and zero-field cooled (ZFC, no magnetic field applied during cooling) protocols were performed. The diamagnetic contribution was systematically subtracted from the silicon substrates.

3. Results

3.1. Morphological and Structural Characterization

One of the aims of this study was to investigate the evolution of the NPs' structure and composition as a function of their size starting from alloyed targets. Initially, two kind of samples were prepared using a $Co_{90}Cr_{10}$ target, varying the Ar flux through the magnetron in use (5 and 10 sccm) and keeping the other fabrication parameters fixed in order to obtain different nanoparticle sizes without a modification of the applied power. Although one of the parameters most used to modify the size of nanoparticles is the applied power, here we have chosen to keep the same power applied to the magnetron in order to avoid differences in the energy supplied to the nanoparticles during their growth, which can condition their final structure [30]. The nanoparticle size was systematically determined by AFM by measuring the height of the nanoparticles from several images, such as the ones depicted in Figure 1a,c and obtaining statistics over 1600 NPs per sample. The resulting height distributions were fitted to a Log-Normal function (Figure 1b,d), as expected using this fabrication method [31], and the extracted mean NP size and their respective standard deviation were found to be 5.59 ± 0.05 nm for the sample fabricated with 10 sccm of Ar and 6.91 ± 0.03 nm when 5 sccm was used, keeping the total Ar flux in the system constant.

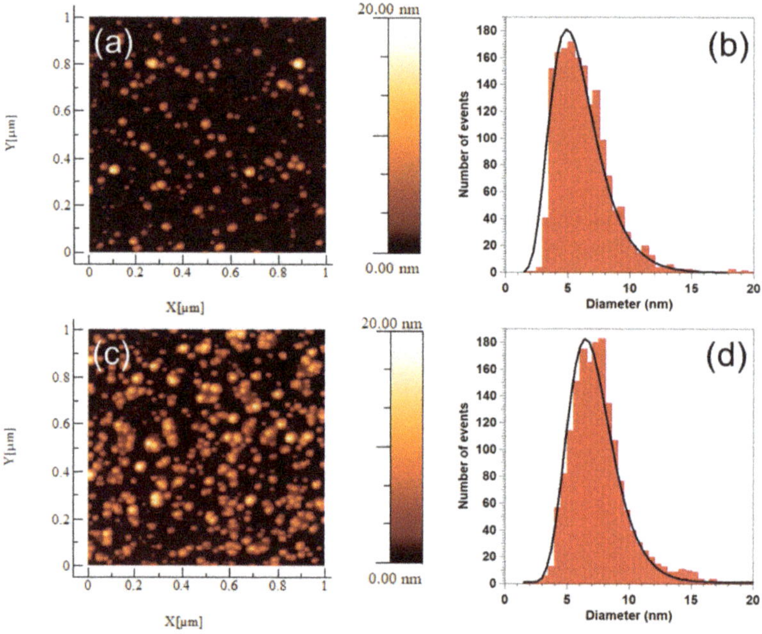

Figure 1. (**a**) AFM image of 1 μm² of a Co@Cr nanoparticle (NP) deposit performed injecting 10 sccm argon into the magnetron loaded with the Co$_{90}$Cr$_{10}$ target, (**b**) height distribution and Log-Normal fitting extracted form (**a**). The mean size of the NPs is ≈5.6 nm, (**c**) the AFM image of 1 μm² of a Co@Cr NP deposit performed with 5 sccm of argon, (**d**) height distribution and Log-Normal fitting extracted form (**c**). The mean size of the NPs is ≈6.9 nm.

Therefore, as can be observed from Figure 1, the modification of the argon flux injected in the working magnetron has a clear influence on the size of the resulting NPs. In previous works using MICS [32,33], we reported how, for the same total Ar flux through the three magnetrons of the MICS, the lower amount of Ar injected through the operating magnetrons leads to bigger NPs.

Spherical aberration-corrected (Cs-corrected) STEM was used for a deeper analysis of the structure and composition of both kinds of NPs. This analysis revealed the dependence of the structure on the NP size. The Cs-corrected high resolution STEM (HR-STEM) images of the smallest NPs (d ≈ 5.6 nm) evidenced the presence of different crystalline domains (Figure 2a), in which EELS analysis revealed a certain Cr segregation (Figure 2b). The homogeneous presence of oxygen indicated that most of Cr and Co were oxidized, although the presence of a small metallic Co core could not be completely discarded. The origin of the oxygen arises from air exposure due to ex situ analysis. On the contrary, the Cs-corrected HR-STEM data of the biggest NPs (d ≈ 6.9 nm) revealed the spontaneous migration of

Cr atoms towards the surface, creating a well-defined shell around a Co core. In this case, the shell could protect the Co core in its metallic state, which explains the contrast difference between the shell and the core, as the former appears with a lower intensity associated with the lower atomic number of oxygen [34]. This shell seems to be formed by small rounded crystallites of 1–2 nm, resulting in a homogenous shell with distinct crystallographic domains in what resembles a petal structure. The EELS analysis revealed that Co is also present in the oxidized shell, but at a lower concentration (Figure 3), and the size of this shell for the $Co_{90}Cr_{10}$ has been measured to be 2.72 ± 0.63 nm.

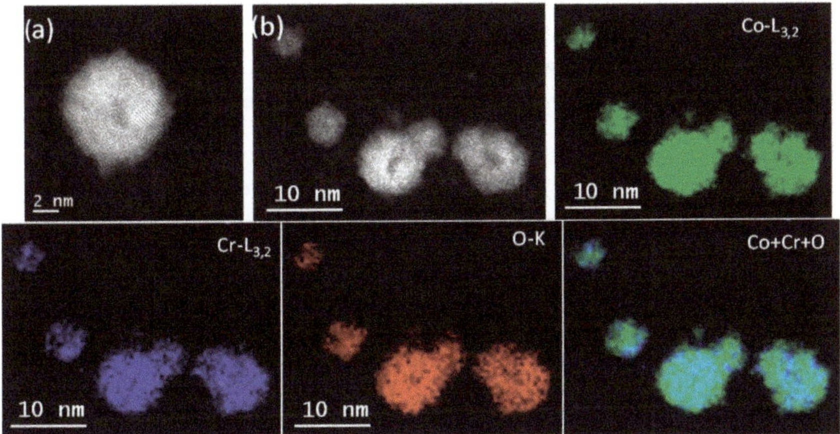

Figure 2. C_s-corrected Scanning High-Resolution Transmission Electron Microscopy (HR-STEM) data corresponding to the small CoCr NPs (5.6 nm). (**a**) High magnification of an isolated NP, exhibiting its different crystal domains. (**b**) Several CoCr NPs and their corresponding Electron Energy Loss Spectroscopy (EELS) chemical maps.

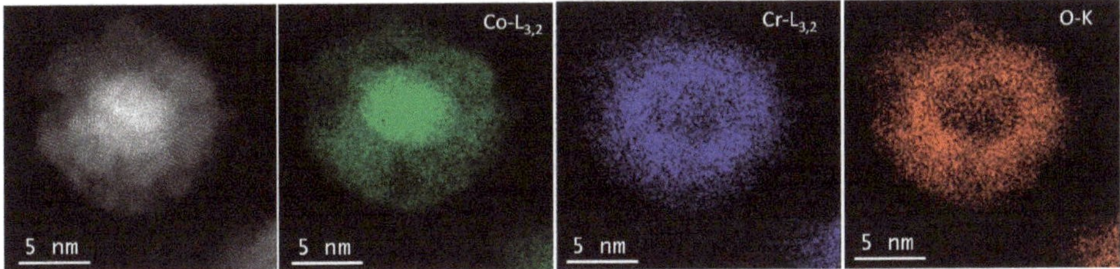

Figure 3. C_s-corrected HR-STEM data corresponding to the bigger CoCr NPs (6.9 nm). A high-magnification image is presented together with the corresponding EELS chemical maps.

According to these results, it seems that a minimum NP size is needed in order to spontaneously generate a core@shell structure without the need of any further treatment. Previous works needed, for instance, a thermal treatment, either in flight [35] or post deposition [36], to create well-defined core@shell NPs. However, the spontaneous formation of this structure clearly simplified the fabrication process. This minimum size is slightly below the 8 nm reported by Koten et al. [37] for similar bimetallic systems. Given that the power used for the fabrication of both NP types was the same, a similar thermal energy should then be expected. Segregation is a size-dependent phenomenon [38], and lowering the free energy for phase separation is significant in this size range [39]. In the case of the smaller NPs, the partial segregation of Cr in certain parts of the surface could suggest that an initial shell formation occurs. Nevertheless, it would not be enough for a complete shell formation, as in the case of the bigger NPs. In addition, the lattice mismatch between both elements implies that segregation would be more favorable than alloy formation, acting in favor of a core@shell configuration. Previous studies have reported the segregation during the gas-phase synthesis of NPs of ferromagnetic elements (Fe, Co) and noble metals (Au, Ag) [40,41]. However, in that case the differences among the characteristics of the two elements, in terms of atomic size and surface energy, were distinctively larger than those between Co and Cr; in fact, they constitute a model system of bimetallic segregated structures [42], as shell formation is strongly favored in that case.

In addition, it is well known that the surface to volume ratio in this size range is highly variable [1]. Assuming that the nanoparticles are spherical, the surface-to-volume ratio of the smaller NPs would be higher than 1 (calculated with the NP radius extracted from the AFM analysis), meaning that there are more atoms at the surface than in the volume (i.e., more than 50% are surface atoms). However, as the percentage of atoms extracted from the sputtering target is only 10% Cr, even taking into account the higher sputtering yield of Cr, there are not enough Cr atoms to create a complete Cr-rich shell covering a Co core, even assuming a complete diffusion of Cr at the surface of the nanoparticles. Nevertheless, the diffusion of the Cr atoms at the NPs surface is observed through the presence of Cr-enriched regions, as shown in Figure 2. It seems then that Cr displays a clear tendency to migrate towards the surface. In the case of bigger NPs (Figure 3), the surface atoms represent only 14% of the total, which is quite close to the 10% Cr atoms available. Therefore, a smaller proportion of atoms are needed to create a Cr-enriched shell in comparison to smaller NPs. It is well known that the thermalization of the nanoparticles takes place through the collisions with the cooling gas (argon, in the case of this work) [43]. Given that the same argon flux was used in the fabrication of both types of nanoparticles, the available argon gas for the cooling process is identical. As the bigger NPs have a higher stored energy (the energy is proportional to the number of atoms in the NP) the cooling process is slower, which favors the diffusion process of the Cr atoms to the surface of the NPs.

3.2. Magnetic Characterization

For the study of the magnetic behavior of the fabricated systems, a single nominal monolayer of NPs (1 ML) was deposited. Consequently, the deposits are subjected to interparticle interaction effects [44]. Dipolar interactions will, on the one hand, increase the blocking temperature [45] and, on the other hand, reduce the coercivity of the loops measured in the blocked regime [46]. There may also be interparticle exchange interactions as a consequence of the connectivity effects between antiferromagnetic shells in contact [47,48].

Figure 4a shows the hysteresis loops registered at 5 K in both samples after cooling in a field H = 50 kOe. The coercivity (H_c) and exchange bias (H_{ex}) fields for the NPs with d = 5.6 nm were found to be 4170 and 1820 Oe, respectively, whereas for the deposit of NPs with d = 6.9 nm, H_c = 3810 Oe and H_{ex} = 1510 Oe were measured.

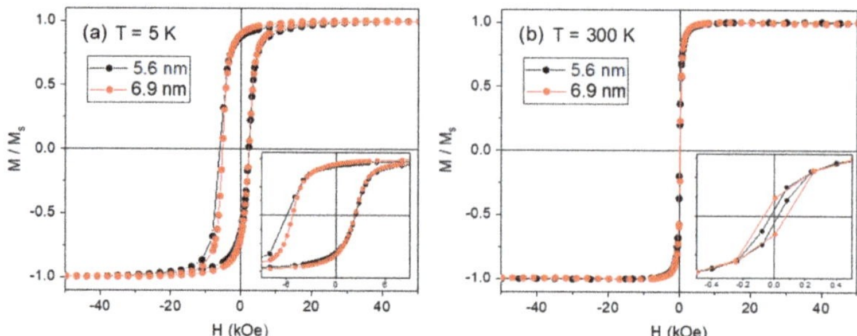

Figure 4. Normalized hysteresis loops measured after cooling in H = 50 kOe at (**a**) 5 K and (**b**) 300 K for CoCr NPs 5.6 nm (black) and 6.9 nm (red) in diameter. The insets show the detail of the central region of the loops, where a strong exchange bias effect can be observed at T = 5 K.

As these small particles are magnetic monodomains [49] well blocked at 5 K, the coercivity should increase with particle size. Yet, the opposite trend is observed. This discrepancy is associated with the inner structure of the NPs unveiled by the above EELS mappings. The bigger NPs present a well-defined core@shell structure with Co as the main constituent in the core, and a surrounding shell comprising Co, Cr, and O. The observation of a coercive field in these NPs indicates that the Co core is metallic. This is probably due to the protective shell, which prevents core oxidation. The smaller NPs present a more heterogeneous structure, where the incomplete Cr shell allows for a higher degree of Co oxidation and, thus, a higher fraction of antiferromagnetic Co oxide, favoring, in turn, a stronger exchange coupling [50,51]. Moreover, for a given antiferromagnetic shell, the interfacial nature of the exchange bias effect yields a dependence on the reciprocal of the magnetic moment of the core [52], also explaining the lower H_{ex} in the larger particles. Exchange coupling is well known to enhance the coercivity of magnetic particles through an increase in the effective anisotropy [52], thus accounting for the observed values.

In contrast, at 300 K (Figure 4b) the H_c of the bigger NPs is slightly higher than that of the smaller NPs. In both cases, exchange bias is reduced to zero, indicating that the phase (or phases) responsible for the exchange coupling must have a Néel temperature, T_N, or a blocking temperature of the antiferromagnetic grains lower than 300 K.

Neither the ZFC nor the FC magnetization curves, measured up to RT, showed a maximum from which a blocking temperature (T_B) could be extracted (see Figure S2, Supplementary Material). This confirms the blocked state at room temperature of our NP systems, with a mean size of 6–7 nm. The ferromagnetic behavior (blocked regime) of such small particles at room temperature (RT) is likely to be associated to the metallic cobalt core assisted by interparticle interactions and by the stabilization of the surface magnetic moments, thanks to the presence of the surrounding phases [53]. The temperature dependence of the coercivity above 150 K is plotted in Figure 5, showing a progressive reduction and vanishing close to 400 K. At 150 K, where the exchange bias field is negligible, the coercivity of the deposit of 6.9 nm NPs is higher than that of 5.6 nm particles, as expected strictly from the particle size dependence of coercivity.

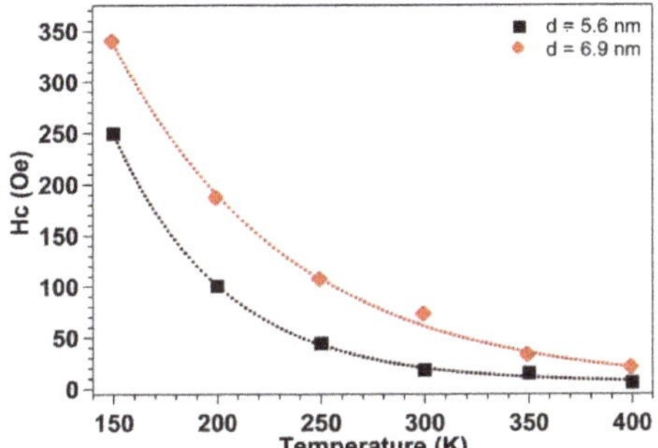

Figure 5. Coercive field dependency on temperature for both samples (5.6 nm, black squares, and 6.9 nm, red diamonds, in diameter).

3.3. Chemical Characterization

In order to extract more information about the chemistry and electronic structure of the core@shell nanoparticles, an XPS analysis was carried out in a multilayer of Co@Cr NPs. A depth profile analysis was performed on the multilayer in order to obtain information on the whole nanostructured system. The sputtering rate is 0.4 Å/μA·min for continuous thin films [54]. Although it is not possible to establish a direct correlation between this value and a porous nanostructured coating, the longest sputtering time used of 375 min ensures that we managed to remove at least one NP layer during the sputtering process. Figure S3 shows the evolution of the main constituents at different sputtering times.

Figure 6a shows the analysis of the Co 2p core-level peak. The fitting was performed using four doublets and was identified using the binding energy (BE) of its $2p_{3/2}$ component. The first doublet at 777.7 ± 0.5 eV corresponds to metallic cobalt [55]. The intensity of this peak is increased with argon sputtering, as expected from a core@shell structure with a metallic cobalt core. The second doublet at 779.9 ± 0.1 eV is usually attributed to cobalt oxides [55]. According to the literature, CoO, Co_2O_3, or Co_3O_4 are present in this BE range [55–57]. Besides, when cobalt is present in Co–Cr spinels ($CoCr_2O_4$) it can be included in this narrow BE range. Consequently, the presence of any of these oxides cannot be ruled out; the coexistence of more than one of them is likely [55,58,59]. However, the presence of a satellite at 786.5 ± 0.1 eV at all depths suggests the presence of Co in the form of CoO among these possible oxide mixtures [59]. Finally, the fourth doublet at 781.2 ± 0.4 eV corresponds to $Co(OH)_2$ [59]. The peak intensity of this component decays exponentially with the sputtering time. Thus, it is considered that this cobalt hydroxide is formed at the surface of the NPs due to atmospheric exposure. A detailed evolution of the metallic, oxide, and hydroxide components as a function of the sputtering time is displayed in Supplementary Figure S4.

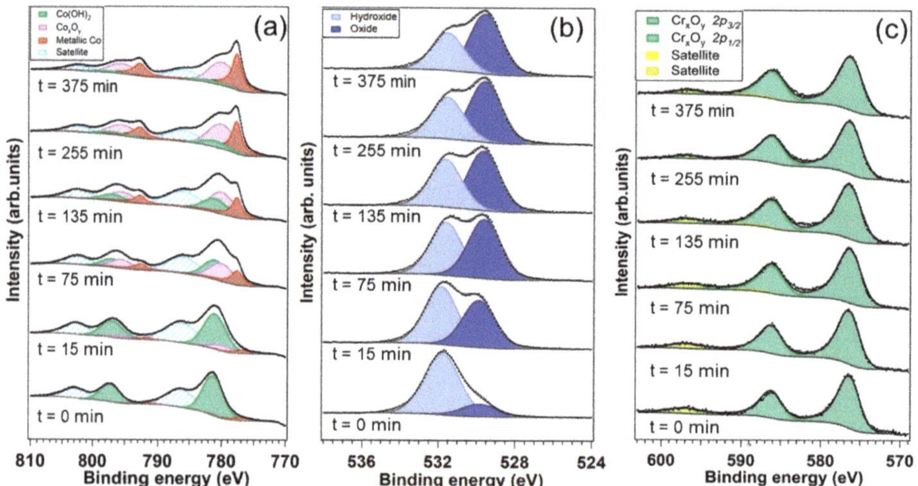

Figure 6. Analysis of the (a) Co 2p, (b) O 1s, and (c) Cr 2p core-level peaks as a function of the sputtering time.

Figure 6b presents the evolution of the O 1s core-level peak with the sputtering time. The spectra were fitted using two components: The first one, at 529.5 ± 0.2 eV, can be associated with metallic oxides. The second, at 531.8 ± 0.2 eV, can be attributed to metallic hydroxides. The compositional ratio of the hydroxide/oxide components in O 1s follows the same tendency as the hydroxide/oxide component evolution in the Co 2p spectra (Figure S4).

Finally, the analysis of the Cr 2p core-level spectra revealed the complete oxidation of chromium, irrespective of the sputtering time (Figure 6c). A single doublet was used for the fitting, with a BE of the $2p_{3/2}$ component at 576.5 ± 0.1 eV, which can be attributed to Cr^{3+} as a constituent of a chromium oxide (Cr_xO_y) or as a part of a mixed oxide Co–Cr ($CoCr_2O_4$). As for the case of the Co 2p, neither of these species can be discarded. The small satellite at 588.5 ± 0.1 eV is characteristic of Cr_2O_3 and, therefore, at least the presence of this oxide can be confirmed. On the contrary, neither the component associated to Cr^0 nor the one of Cr^{6+} could be detected [55].

In summary, the XPS analyses are consistent with a core@shell structure in which a metallic Co core is surrounded by a shell composed of Co and Cr oxides and a surface enriched in $Co(OH)_2$ due to atmospheric exposure. Both elements, Cr and Co, have a strong tendency to become oxidized.

According to the XPS analysis, Cr is mainly found as Cr^{3+} and Co as Co^{2+}, which means that Cr has one unpaired d electron more and, therefore, a higher tendency to get oxidized. Considering the fact that Cr is mainly located on the surface, and in form of chromium oxide, the possibility of the formation of a passive layer as the one Cr forms on stainless steels cannot be ruled out. Similar behavior has been reported on CoCr alloys with higher Cr contents, where a preferential Cr oxidation was observed in a surface enriched in Cr with Co depletion [60]. This is compatible with the presence of a metallic Co core inside the NPs.

3.4. Nanoparticles with a Different Stoichiometry

According to the previous sections, one strategy to obtain a core@shell structure consists of decreasing the ratio between surface and bulk atoms, while keeping fixed the ratio of Co and Cr atoms in the sputtering target. Nevertheless, an alternative and more common strategy consists of increasing the content of shell (Cr) atoms. This latter approach has been adopted to fabricate NPs where a sputtering target with a higher content of Cr ($Co_{80}Cr_{20}$) has been used. The growth parameters were adjusted in order to obtain nanoparticles with a mean diameter close to 7 nm in order to allow the comparison with the previous Co@Cr NPs grown with the $Co_{90}Cr_{10}$ target.

Figure 7 shows a representative C_s-corrected HR-STEM image of a typical NP obtained in this new target. As can be observed, also in this case the core@shell structure is the most stable configuration. The EELS analysis revealed that Co is located within the NP core, while Cr is forming the shell, in the same way as for the previous NPs grown with the lower argon flux and the $Co_{90}Cr_{10}$ target. The presence of oxygen is also evidenced, preferentially where Cr is present, due to air exposure. From a statistical analysis of the TEM images, an average NP size of 7.7 ± 0.5 nm, with an average core 4.3 ± 0.3 nm, and a shell 1.70 ± 0.11 nm can be extracted. Taking into account the target composition ($Co_{80}Cr_{20}$) and considering that the core is formed by pure Co, a core@shell structure should have a 5.4 nm diameter core with a shell of 0.55 nm, i.e., three times smaller than the obtained shell. This mismatch could arise from the fact that Cr is oxidized and the oxide has a bigger lattice parameter than pure Cr [61]. Nevertheless, a certain presence of Co in the shell cannot be completely ruled out.

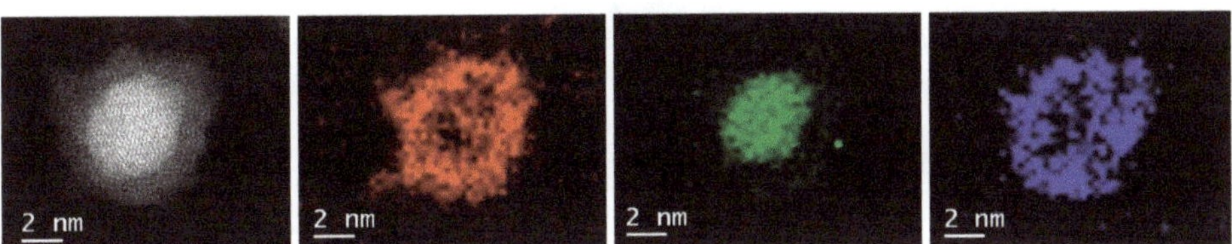

Figure 7. Cs-corrected HR-STEM data corresponding to the CoCr NPs from the $Co_{80}Cr_{20}$ target. A high-magnification image is presented together with the corresponding EELS chemical maps.

Line profiles were also extracted from the three samples (see Figure 8). Figure 8a corresponds to the smallest $Co_{90}Cr_{10}$ NPs displaying a homogeneous metal distribution (this is no core@shell structures). For the samples with a diameter of 6.9 nm (Figure 8b,c for $Co_{90}Cr_{10}$ and $Co_{80}Cr_{20}$, respectively) the core@shell structure is evidenced. For $Co_{90}Cr_{10}$, the Co, although mainly centered in the core, is also clearly present in the shell, suggesting that this shell is formed by Co, Cr, and O. On the other hand, for $Co_{80}Cr_{20}$ material the shell is mainly formed by strongly oxidized Cr. The fact the outermost shell is formed by the three compounds for the Co-rich material may explain the larger size of the shell.

The power used for the NP fabrication (5 W) was nearly three times lower than in the previous case (14 W). Interestingly, despite this lower applied power to the magnetron, 7 nm diameter nanoparticles have been obtained. The reason for this could be understood in terms of the higher sputtering yield of Cr, as well as its larger content in the target. In any case, this lower power implies that a smaller thermal energy supplied to the NPs during the fabrication process. However, a thicker shell was systematically observed in the TEM analysis of these NPs (see Supplementary Figure S5) and revealed the fact that Cr diffusion is again highly favored.

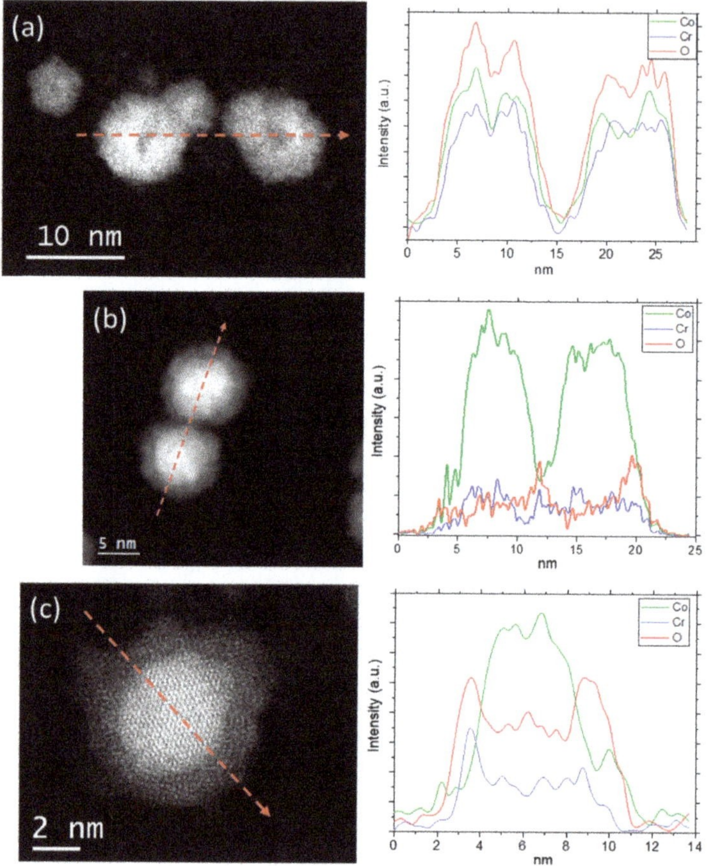

Figure 8. EELS profiles extracted along the red dash arrows corresponding to (**a**) $Co_{90}Cr_{10}$ 5.6 nm, (**b**) $Co_{90}Cr_{10}$ 6.9 nm, and (**c**) $Co_{80}Cr_{20}$ 7 nm.

The well-defined core@shell structure gives us the opportunity to try to access the fundamental magnetic properties of the NPs. Hence, in this case, the magnetic characterization was carried out on samples with less than a monolayer of nanoparticles. This allowed us to reduce the interparticle interaction, leading to a deep understanding of the magnetic properties of the NPs themselves. Nevertheless, a minimal density is needed to gain access to some magnetic signal. The compromise was found in samples with 15% of a monolayer of nanoparticles.

Figure 9a shows ZCF-FC curves, where the diamagnetic contribution of the Si substrate ($M_{Si} = 5.2 \times 10^{-6}$ emu) has been corrected. In this system, a blocking temperature of $T_B = 70 \pm 10$ K was found (cf. Figure S6), a value that is even higher than reported $T_B = 50$ K for Co NPs of 10 nm in diameter [44]. In fact, the T_B of these Co@Cr NPs is similar to that expected for Co NPs with a diameter twice as large, which implies a corresponding enhancement of the effective anisotropy in our Co@Cr NPs. Since, for the given surface coverage, dipolar interparticle interactions are expected to have only a moderate effect on the blocking temperature [62], this result is attributed to exchange coupling between the ferromagnetic Co core and an antiferromagnetic shell.

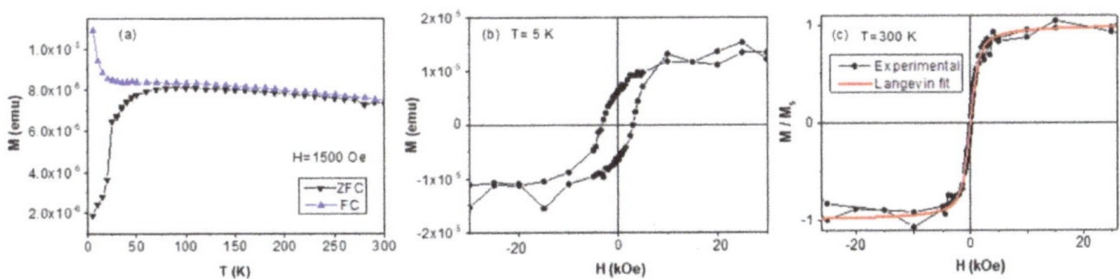

Figure 9. (**a**) Zero-field cooled-field-cooled (ZFC-FC) curves measured at H = 1500 Oe. Hysteresis loop measured after zero-field cooling at (**b**) 5 K and (**c**) 300 K.

The hysteresis loop displayed in Figure 9b, measured at 5 K, shows a coercive field of $H_c = (3.2 \pm 0.3)$ kOe, approximately 16% lower than that measured for the Co@Cr NPs produced with the $Co_{90}Cr_{10}$ target (Figure 4). The difference can be attributed to both a smaller Co core and a weaker exchange coupling (due to a reduced shell oxidation) in the case of the $Co_{80}Cr_{20}$ system.

In contrast, the hysteresis loop at 300 K (Figure 9c) shows negligible remanence and coercivity, as expected for the superparamagnetic (SP) response indicated by Figure 9a. This SP regime is, in turn, expected from the small diameter estimated above for the Co core ($d_{TEM} \approx 4.3$ nm in diameter), since the critical SP size at RT for a Co NP is $d \approx 9$ nm [63]. The magnetic volume (V_{mag}) of the sample can be extracted by fitting the observed SP response to the Langevin function [64,65], obtaining $V_{mag} = 5.3 \pm 0.3 \times 10^{-20}$ cm^3, from which the *magnetic* diameter $d_{mag} = 4.7 \pm 0.2$ nm follows, in good agreement with d_{TEM}, particularly after taking into account that ignoring the size distribution in a Langevin fit will yield an effective particle moment larger than the mean particle moment of the distribution [62]. Note that the use of the Langevin equation, which assumes a negligible anisotropy barrier compared to the thermal energy, is justified in the first place by the fact that the measurement temperature (300 K) is more than four times higher than the blocking temperature of any of the samples, the criterion suggested by Mamiya et al. for typical size distributions [66].

Once the size of the Co core has been established, the effective magnetic anisotropy constant of the Co@Cr NPs ($K_{Co@Cr}$) can be extracted from the Néel–Brown law governing the relaxation time of the SP NPs as [13]:

$$K_{Co@Cr} = \frac{25 \times T_B \times k_B}{\frac{\pi}{6} \times d^3} = (5.8 \pm 1.2) \times 10^5 \, \text{J/m}^3. \tag{1}$$

The value obtained nearly doubles the volume anisotropy for FCC Co ($K_{Co} = 2.7 \times 10^5$ J/m^3) [67]. As mentioned above, this enhanced value could possibly be due to the existence of an additional contribution to the anisotropy from the exchange coupling between the ferromagnetic core and an antiferromagnetic shell.

4. Conclusions

Well-defined Core@shell Co@Cr nanoparticles can be spontaneously formed in the gas phase under certain experimental conditions using alloy targets. Starting from a $Co_{90}Cr_{10}$ target, we observe that a minimum NP size is needed to obtain shell formation without further treatments. When the number of atoms at the surface is higher than number of atoms at the volume, there is not enough Cr to create a shell and only certain Cr segregation is observed. This occurred for the NPs with a mean diameter of 5.6 nm. On the other hand, in the NPs with a mean diameter of 6.9 nm the number of surface atoms becomes smaller than the volume atoms and a core@shell structure is formed. In this system, a Cr-enriched shell surrounds a Co core. For a similar NP size, when the Cr content is increased (using a $Co_{80}Cr_{20}$ target) a thicker Cr coating is formed, as expected from the higher Cr content. These results evidenced the tendency of Cr to migrate towards the surface to form a shell, which subsequently oxidized after air exposure.

In these conditions, ferromagnetic NPs stable in a wide temperature range (below 300 K) are obtained using a $Co_{90}Cr_{10}$ target, reducing the critical size for room temperature superparamagnetism to well below 7 nm. This is caused by the enhancement of the total anisotropy due to an exchange bias contribution stemming from the coupling between the different magnetic phases. In the smaller NPs (5.6 nm), there is a larger presence of interfaces due to the more heterogeneous structure, which favors an increase in the coercive field and the exchange bias. This, together with the dipolar interactions between NPs, is responsible for the hard magnetism at low temperatures and the blocked response at RT. Although the NPs with a larger proportion of Cr remained superparamagnetic at RT, they present nearly twice as much anisotropy as conventional FCC Co NPs.

Supplementary Materials: The following are available online at http://www.mdpi.com/2673-3501/1/1/7/s1: Figure S1. Scheme of the MICS system for the fabrication of the nanoparticles; Figure S2: ZFC-FC magnetization curves of (a) NPs d = 5.6 nm at H_{apl} = 100 Oe and (b) NPs of d = 6.9 nm at H_{apl} = 50 Oe; Figure S3: Evolution of the chemical composition extracted from the wide scan spectra taken at different sputtering times with Ar^+; Figure S4: Evolution of the cobalt components with sputtering time; Figure S5: Comparison of the NPs formed from a $Co_{90}Cr_{10}$ and $Co_{80}Cr_{20}$ target; Figure S6: log (MFC-MZFC) vs. T of the $Co_{80}Cr_{20}$ 7 nm NPs. A change in the slope can be observed at T $\approx$ 70 K, indicating the blocking temperature.

Author Contributions: Conceptualization, Y.H.; methodology, L.M., E.N., and Y.H.; data acquisition J.S.-M., E.M.J., L.M., Á.M., E.H.S., J.A.D.T., E.N., Y.H.; formal analysis J.S.-M., E.M.J., L.M., Á.M., E.H.S., J.A.D.T., E.N., Y.H.; investigation, J.S.-M. and E.M.J.; writing—original draft preparation, L.M.; writing—review and editing J.S.-M., E.M.J., L.M., Á.M., E.H.S., J.A.D.T., E.N., Y.H.; supervision, Y.H. All authors have read and agreed to the published version of the manuscript.

Funding: This research was funded by the Spanish Ministry of Science, grant number RYC2018-024561-I; the National Natural Science Foundation of China, Grant Number NFSC-21850410448; The Centre for High-resolution Electron Microscopy (CℏEM), supported by SPST of ShanghaiTech University under contract No. EM02161943 and NSFC- 21835002; the Spanish MINECO, grant number FIS2016-76058-C4-1-R and MAT2015-65295-R; and the Comunidad de Madrid grant number P2018/NMT-4321.

Conflicts of Interest: The authors declare no conflict of interest.

References

1. He, L.-B.; Zhang, L.; Tang, L.-P.; Sun, J.; Zhang, Q.-B.; Sun, L.-T. Novel behaviors/properties of nanometals induced by surface effects. *Mater. Today Nano* **2018**, *1*, 8–21. [CrossRef]

2. González-Rubio, G.; Mosquera, J.; Kumar, V.; Pedrazo-Tardajos, A.; Llombart, P.; Solís, D.M.; Lobato, I.; Noya, E.G.; Guerrero-Martínez, A.; Taboada, J.M.; et al. Micelle-directed chiral seeded growth on anisotropic gold nanocrystals. *Science* **2020**, *368*, 1472–1477. [CrossRef]

3. Seh, Z.W.; Kibsgaard, J.; Dickens, C.F.; Chorkendorff, I.; Nørskov, J.K.; Jaramillo, T.F. Combining theory and experiment in electrocatalysis: Insights into materials design. *Science* **2017**, *355*, eaad4998. [CrossRef] [PubMed]

4. Qiu, L.; Zhu, N.; Feng, Y.; Michaelides, E.E.; Żyła, G.; Jing, D.; Zhang, X.; Norris, P.M.; Markides, C.N.; Mahian, O. A review of recent advances in thermophysical properties at the nanoscale: From solid state to colloids. *Phys. Rep.* **2020**, *843*, 1–81. [CrossRef]

5. Jun, Y.-W.; Seo, J.-W.; Cheon, J. Nanoscaling Laws of Magnetic Nanoparticles and Their Applicabilities in Biomedical Sciences. *Acc. Chem. Res.* **2008**, *41*, 179–189. [CrossRef]

6. Binns, C.; Trohidou, K.N.; Bansmann, J.; Baker, S.H.; A Blackman, J.; Bucher, J.-P.; Kechrakos, D.; Kleibert, A.; Louch, S.; Meiwes-Broer, K.-H.; et al. The behaviour of nanostructured magnetic materials produced by depositing gas-phase nanoparticles. *J. Phys. D Appl. Phys.* **2005**, *38*, R357–R379. [CrossRef]

7. De Toro, J.A.; Vasilakaki, M.; Lee, S.S.; Andersson, M.S.; Normile, P.S.; Yaacoub, N.; Murray, P.D.; Sánchez, E.H.; Muñiz, P.; Peddis, D.; et al. Remanence Plots as a Probe of Spin Disorder in Magnetic Nanoparticles. *Chem. Mater.* **2017**, *29*, 8258–8268. [CrossRef]

8. Bucher, J.P.; Douglass, D.C.; Bloomfield, L.A. Magnetic properties of free cobalt clusters. *Phys. Rev. Lett.* **1991**, *66*, 3052–3055. [CrossRef]

9. Suzuki, M.; Muraoka, H.; Inaba, Y.; Miyagawa, H.; Kawamura, N.; Shimatsu, T.; Maruyama, H.; Ishimatsu, N.; Isohama, Y.; Sonobe, Y. Depth profile of spin and orbital magnetic moments in a subnanometer Pt film on Co. *Phys. Rev. B* **2005**, *72*, 54430. [CrossRef]

10. Edmonds, K.W.; Binns, C.; Baker, S.; Maher, M.; Thornton, S.; Tjernberg, O.; Brookes, N.B. Magnetism of exposed and Co-capped Fe nanoparticles. *J. Magn. Magn. Mater.* **2000**, *220*, 25–30. [CrossRef]

11. Baker, S.; Binns, C.; Edmonds, K.W.; Maher, M.; Thornton, S.; Louch, S.; Dhesi, S. Enhancements in magnetic moments of exposed and Co-coated Fe nanoclusters as a function of cluster size. *J. Magn. Magn. Mater.* **2002**, *247*, 19–25. [CrossRef]

12. Popok, V.; Kylian, O. Gas-Phase Synthesis of Functional Nanomaterials. *Appl. Nano* **2020**, *1*, 25–58. [CrossRef]

13. Blundell, S. *Magnetism in Condensed Matter*; Oxford University Press: Oxford, UK, 2001; ISBN 9780198505914.

14. Baibich, M.N.; Broto, J.M.; Fert, A.; Van Dau, F.N.; Petroff, F.; Etienne, P.; Creuzet, G.; Friederich, A.; Chazelas, J. Giant Magnetoresistance of (001)Fe/(001)Cr Magnetic Superlattices. *Phys. Rev. Lett.* **1988**, *61*, 2472–2475. [CrossRef]

15. Binasch, G.; Grünberg, P.; Saurenbach, F.; Zinn, W. Enhanced magnetoresistance in layered magnetic structures with antiferromagnetic interlayer exchange. *Phys. Rev. B* **1989**, *39*, 4828–4830. [CrossRef]

16. Skumryev, V.; Stoyanov, S.; Zhang, Y.; Hadjipanayis, G.; Givord, D.; Nogués, J. Beating the superparamagnetic limit with exchange bias. *Nat. Cell Biol.* **2003**, *423*, 850–853. [CrossRef]

17. Sato, N. Structure and magnetism of Co-Cr thin films with an artificial superlattice structure. *J. Appl. Phys.* **1987**, *61*, 1979–1989. [CrossRef]

18. Morikawa, T.; Suzuki, M.; Taga, Y. Soft magnetic properties of Co–Cr–O granular films. *J. Appl. Phys.* **1998**, *83*, 6664–6666. [CrossRef]

19. Fallarino, L.; Kirby, B.J.; Pancaldi, M.; Riego, P.; Balk, A.L.; Miller, C.W.; Vavassori, P.; Berger, A. Magnetic properties of epitaxial CoCr films with depth-dependent exchange-coupling profiles. *Phys. Rev. B* **2017**, *95*, 134445. [CrossRef]

20. Öztürk, M.; Demirci, E.; Erkovan, M.; Öztürk, O.; Akdoğan, N. Coexistence of perpendicular and in-plane exchange bias using a single ferromagnetic layer in Pt/Co/Cr/CoO thin film. *EPL Europhys. Lett.* **2016**, *114*, 17008. [CrossRef]

21. Cui, B.; Li, D.; Yun, J.; Zuo, Y.; Guo, X.; Wu, K.; Zhang, X.; Wang, Y.; Xi, L.; Xue, D. Magnetization switching through domain wall motion in Pt/Co/Cr racetracks with the assistance of the accompanying Joule heating effect. *Phys. Chem. Chem. Phys.* **2018**, *20*, 9904–9909. [CrossRef] [PubMed]

22. Petr, M.; Kylián, O.; Kuzminova, A.; Kratochvíl, J.; Khalakhan, I.; Hanuš, J.; Biederman, H. Noble metal nanostructures for double plasmon resonance with tunable properties. *Opt. Mater.* **2017**, *64*, 276–281. [CrossRef]

23. Zhao, J.; Mayoral, A.; Martínez, L.; Johansson, M.P.; Djurabekova, F.; Huttel, Y. Core–Satellite Gold Nanoparticle Complexes Grown by Inert Gas-Phase Condensation. *J. Phys. Chem. C* **2020**. [CrossRef] [PubMed]

24. Haberland, H.; Karrais, M.; Mall, M. A new type of cluster and cluster ion source. *Eur. Phys. J. D* **1991**, *20*, 413–415. [CrossRef]

25. Martínez, L.; Díaz, M.; Román, E.; Ruano, M.; P., D.L.; Huttel, Y. Generation of Nanoparticles with Adjustable Size and Controlled Stoichiometry: Recent Advances. *Langmuir* **2012**, *28*, 11241–11249. [CrossRef]

26. Llamosa, D.; Ruano, M.; Martínez, L.; Mayoral, A.; Roman, E.; García-Hernández, M.; Huttel, Y. The ultimate step towards a tailored engineering of core@shell and core@shell@shell nanoparticles. *Nanoscale* **2014**, *6*, 13483–13486. [CrossRef] [PubMed]

27. Spadaro, M.C.; Humphrey, J.J.L.; Cai, R.; Martínez, L.; Haigh, S.J.; Huttel, Y.; Spencer, S.J.; Wain, A.J.; Palmer, R. Electrocatalytic Behavior of PtCu Clusters Produced by Nanoparticle Beam Deposition. *J. Phys. Chem. C* **2020**. [CrossRef] [PubMed]

28. Horcas, I.; Fernández, R.; Gómez-Rodríguez, J.M.; Colchero, J.; Gómez-Herrero, J.; Baro, A.M. WSXM: A software for scanning probe microscopy and a tool for nanotechnology. *Rev. Sci. Instrum.* **2007**, *78*, 013705. [CrossRef]

29. Briggs, D. Handbook of X-ray Photoelectron Spectroscopy C. D. Wanger, W. M. Riggs, L. E. Davis, J. F. Moulder and G. E.Muilenberg Perkin-Elmer Corp., Physical Electronics Division, Eden Prairie, Minnesota, USA, 1979. 190 pp. $195. *Surf. Interface Anal.* **1981**, *3*. [CrossRef]

30. Krishnan, G.; Verheijen, M.A.; Brink, G.H.T.; Palasantzas, G.; Kooi, B.J. Tuning structural motifs and alloying of bulk immiscible Mo–Cu bimetallic nanoparticles by gas-phase synthesis. *Nanoscale* **2013**, *5*, 5375–5383. [CrossRef]

31. Shyjumon, I.; Gopinadhan, M.; Helm, C.A.; Smirnov, B.M.; Hippler, R. Deposition of titanium/titanium oxide clusters produced by magnetron sputtering. *Thin Solid Films* **2006**, *500*, 41–51. [CrossRef]

32. Perez, D.L.; Martínez, L.; Huttel, Y.; Nez, L. Martí Multiple Ion Cluster Source for the Generation of Magnetic Nanoparticles: Investigation of the Efficiency as a Function of the Working Parameters for the Case of Cobalt. *Dataset Pap. Sci.* **2014**, *2014*, 1–9. [CrossRef]

33. Ruano, M.; Huttel, Y.; Nez, L. Martí Investigation of the Working Parameters of a Single Magnetron of a Multiple Ion Cluster Source: Determination of the Relative Influence of the Parameters on the Size and Density of Nanoparticles. *Dataset Pap. Sci.* **2013**, *2013*, 1–8. [CrossRef]

34. Mayoral, A.; Mejía-Rosales, S.; Mariscal, M.M.; Pérez-Tijerina, E.; Jose-Yacaman, M. The Co–Au interface in bimetallic nanoparticles: A high resolution STEM study. *Nanoscale* **2010**, *2*, 2647. [CrossRef]

35. Binns, C.; Prieto, P.; Baker, S.; Howes, P.; Dondi, R.; A Burley, G.; Lari, L.; Kröger, R.; Pratt, A.; Aktas, S.; et al. Preparation of hydrosol suspensions of elemental and core–shell nanoparticles by co-deposition with water vapour from the gas-phase in ultra-high vacuum conditions. *J. Nanoparticle Res.* **2012**, *14*, 1–16. [CrossRef]

36. Pérez, D.L.; Espinosa, A.; Martínez-Orellana, L.; Román, E.; Ballesteros, C.; Mayoral, A.; García, E.L.R.; Huttel, Y. Thermal Diffusion at Nanoscale: From CoAu Alloy Nanoparticles to Co@Au Core/Shell Structures. *J. Phys. Chem. C* **2013**, *117*, 3101–3108. [CrossRef]

37. Koten, M.A.; Mukherjee, P.; Shield, J.E. Core-Shell Nanoparticles Driven by Surface Energy Differences in the Co-Ag, W-Fe, and Mo-Co Systems. *Part. Part. Syst. Charact.* **2015**, *32*, 848–853. [CrossRef]

38. Peng, L.; Ringe, E.; Van Duyne, R.P.; Marks, L. Segregation in bimetallic nanoparticles. *Phys. Chem. Chem. Phys.* **2015**, *17*, 27940–27951. [CrossRef]

39. Palomares-Baez, J.-P.; Panizon, E.; Ferrando, R. Nanoscale Effects on Phase Separation. *Nano Lett.* **2017**, *17*, 5394–5401. [CrossRef]

40. Xu, Y.-H.; Wang, J.-P. Direct Gas-Phase Synthesis of Heterostructured Nanoparticles through Phase Separation and Surface Segregation. *Adv. Mater.* **2008**, *20*, 994–999. [CrossRef]

41. Xu, Y.-H.; Wang, J.-P. Magnetic Properties of Heterostructured Co–Au Nanoparticles Direct-Synthesized from Gas Phase. *IEEE Trans. Magn.* **2007**, *43*, 3109–3111. [CrossRef]

42. Vernieres, J.; Steinhauer, S.; Zhao, J.; Chapelle, A.; Menini, P.; Dufour, N.; Diaz, R.E.; Nordlund, K.; Djurabekova, F.; Grammatikopoulos, P.; et al. Gas Phase Synthesis of Multifunctional Fe-Based Nanocubes. *Adv. Funct. Mater.* **2017**, *27*, 1605328. [CrossRef]

43. Kashtanov, P.V.; Smirnov, B.M.; Hippler, R. Magnetron plasma and nanotechnology. *Phys. Uspekhi* **2007**, *50*, 455. [CrossRef]

44. Ruano, M.; Diaz, M.; Martinez, L.; Navarro, E.; Roman, E.; Garcia-Hernandez, M.; Espinosa, A.; Ballesteros, C.I.; Fermento, R.; Huttel, Y. Matrix and interaction effects on the magnetic properties of Co nanoparticles embedded in gold and vanadium. *Phys. Chem. Chem. Phys.* **2013**, *15*, 316–329. [CrossRef] [PubMed]

45. Andersson, M.S.; Mathieu, R.; Lee, S.S.; Normile, P.S.; Singh, G.; Nordblad, P.; De Toro, J.A. Size-dependent surface effects in maghemite nanoparticles and its impact on interparticle interactions in dense assemblies. *Nanotechnology* **2015**, *26*, 475703. [CrossRef]

46. Sanchez, E.H.; Vasilakaki, M.; Lee, S.S.; Normile, P.S.; Muscas, G.; Murgia, M.; Andersson, M.S.; Singh, G.; Mathieu, R.; Nordblad, P.; et al. Simultaneous Individual and Dipolar Collective Properties in Binary Assemblies of Magnetic Nanoparticles. *Chem. Mater.* **2020**, *32*, 969–981. [CrossRef]

47. López-Ortega, A.; Estrader, M.; Salazar-Alvarez, G.; Roca, A.G.; Nogués, J. Applications of exchange coupled bi-magnetic hard/soft and soft/hard magnetic core/shell nanoparticles. *Phys. Rep.* **2015**, *553*, 1–32. [CrossRef]

48. Nogués, J.; Skumryev, V.; Sort, J.; Stoyanov, S.; Givord, D. Shell-Driven Magnetic Stability in Core-Shell Nanoparticles. *Phys. Rev. Lett.* **2006**, *97*, 157203. [CrossRef]

49. Coey, J.M.D. *Magnetism and Magnetic Materials*; Cambridge University Press: Cambridge, UK, 2010.

50. De Toro, J.A.; Andrés, J.P.; Gonzalez, J.A.; Muñiz, P.; Riveiro, J.M. The oxidation of metal-capped Co cluster films under ambient conditions. *Nanotechnology* **2009**, *20*, 085710. [CrossRef]

51. Normile, P.S.; De Toro, J.A.; Muñoz, T.; Gonzalez, J.A.; Andrés, J.P.; Muñiz, P.; Escobar-Galindo, R.; Riveiro, J.M. Influence of spacer layer morphology on the exchange-bias properties of reactively sputteredCo/Agmultilayers. *Phys. Rev. B* **2007**, *76*, 104430. [CrossRef]

52. Nogues, J.; Sort, J.; Langlais, V.; Skumryev, V.; Suriñach, S.; Muñoz, J.; Baró, M. Exchange bias in nanostructures. *Phys. Rep.* **2005**, *422*, 65–117. [CrossRef]

53. Kalska-Szostko, B.; Fumagalli, P.; Hilgendorff, M.; Giersig, M. Co/CoO core–shell nanoparticles—Temperature-dependent magneto-optic studies. *Mater. Chem. Phys.* **2008**, *112*, 1129–1132. [CrossRef]

54. Martín-González, M.; Huttel, Y.; Cebollada, A.; Reig, G.A.; Briones, F. Surface localized nitrogen incorporation in epitaxial FePd films and its effect in the Magneto-Optical properties. *Surf. Sci.* **2004**, *571*, 63–73. [CrossRef]

55. Surviliene, S. The use of XPS for study of the surface layers of Cr–Co alloy electrodeposited from Cr(III) formate–urea baths. *Solid State Ion.* **2008**, *179*, 222–227. [CrossRef]

56. Dupin, J.-C.; Gonbeau, D.; Vinatier, P.; Levasseur, A. Systematic XPS studies of metal oxides, hydroxides and peroxides. *Phys. Chem. Chem. Phys.* **2000**, *2*, 1319–1324. [CrossRef]

57. La Rosa-Toro, A.; Berenguer, R.; Quijada, C.; Montilla, F.; Morallón, E.; Vázquez, J.L. Preparation and Characterization of Copper-Doped Cobalt Oxide Electrodes. *J. Phys. Chem. B* **2006**, *110*, 24021–24029. [CrossRef]

58. Hosseini, S.A.; Alvarez-Galvan, M.; Fierro, J.; Niaei, A.; Salari, D. MCr2O4 (M=Co, Cu, and Zn) nanospinels for 2-propanol combustion: Correlation of structural properties with catalytic performance and stability. *Ceram. Int.* **2013**, *39*, 9253–9261. [CrossRef]

59. Rosa, W.O.; Martínez, L.; Jaafar, M.; Asenjo, A.; Vázquez, M. Asymmetric magnetization reversal process in Co nanohill arrays. *J. Appl. Phys.* **2009**, *106*, 103906. [CrossRef]

60. Tsustumi, Y.; Doi, H.; Nomura, N.; Ashida, M.; Chen, P.; Kawasaki, A.; Hanawa, T. Surface Composition and Corrosion Resistance of Co-Cr Alloys Containing High Chromium. *Mater. Trans.* **2016**, *57*, 2033–2040. [CrossRef]

61. Abdullah, M.M.; Rajab, F.M.; Al-Abbas, S.M. Structural and optical characterization of Cr2O3 nanostructures: Evaluation of its dielectric properties. *AIP Adv.* **2014**, *4*, 027121. [CrossRef]

62. De Toro, J.A.; Gonzalez, J.A.; Normile, P.S.; Muñiz, P.; Andrés, J.P.; Antón, R.L.; Canales-Vázquez, J.; Riveiro, J.M. Energy barrier enhancement by weak magnetic interactions in Co/Nb granular films assembled by inert gas condensation. *Phys. Rev. B* **2012**, *85*, 054429. [CrossRef]

63. Llamosa Pérez, D. *Fabricación y Estudio de las Propiedades Físicas de Nanopartículas de Aleación, Núcleo@corteza y Núcleo@corteza@corteza Basadas en Co, Au y Ag*; Universidad Autónoma de Madrid: Madrid, Spain, 2014.

64. Shiratsuchi, Y.; Yamamoto, M.; Endo, Y.; Li, D.; Bader, S.D. Superparamagnetic behavior of ultrathin Fe films grown on Al$_2$O$_3$(0001) substrates. *J. Appl. Phys.* **2003**, *94*, 7675–7679. [CrossRef]

65. Tamion, A.; Hillenkamp, M.; Tournus, F.; Bonet, E.; Dupuis, V. Accurate determination of the magnetic anisotropy in cluster-assembled nanostructures. *Appl. Phys. Lett.* **2009**, *95*, 62503. [CrossRef]

66. Mamiya, H.; Nakatani, I.; Furubayashi, T.; Ohnuma, M. Analyses of Superparamagnetism-Magnetic Properties of Isolated Iron -Nitride Nanoparticles. *Trans. Magn. Soc. Jpn.* **2002**, *2*, 36–48. [CrossRef]

67. Chen, J.-P.; Sorensen, C.M.; Klabunde, K.J.; Hadjipanayis, G.C. Magnetic properties of nanophase cobalt particles synthesized in inversed micelles. *J. Appl. Phys.* **1994**, *76*, 6316–6318. [CrossRef]

Publisher's Note: MDPI stays neutral with regard to jurisdictional claims in published maps and institutional affiliations.

 *applied nano*

Article

Fabrication of High-Aspect-Ratio Cylindrical Micro-Structures Based on Electroactive Ionogel/Gold Nanocomposite

Edoardo Milana [1,†], Tommaso Santaniello [2,*,†], Paolo Azzini [2], Lorenzo Migliorini [2] and Paolo Milani [2]

[1] Department of Mechanical Engineering, KU Leuven and Flanders Make, Celestijnenlaan 300, 3001 Leuven, Belgium; edoardo.milana@kuleuven.be
[2] CIMaINa, Department of Physics, Università degli Studi di Milano, Via Celoria 16, 20133 Milano, Italy; paolo.azzini@studenti.unimi.it (P.A.); lorenzo.migliorini@unimi.it (L.M.); paolo.milani@mi.infn.it (P.M.)
* Correspondence: tommaso.santaniello@unimi.it
† The authors equally contributed to the work.

Received: 29 September 2020; Accepted: 21 October 2020; Published: 26 October 2020

Abstract: We present a fabrication process to realize 3D high-aspect-ratio cylindrical micro-structures of soft ionogel/gold nanocomposites by combining replica molding and Supersonic Cluster Beam Deposition (SCBD). Cylinders' metallic masters (0.5 mm in diameter) are used to fabricate polydimethylsiloxane (PDMS) molds, where the ionogel is casted and UV cured. The replicated ionogel cylinders (aspect ratio > 20) are subsequently metallized through SCBD to integrate nanostructured gold electrodes (150 nm thick) into the polymer. Nanocomposite thin films are characterized in terms of electrochemical properties, exhibiting large double layer capacitance (24 μF/cm^2) and suitable ionic conductivity (0.05 mS/cm) for charge transport across the network. Preliminary actuation tests show that the nanocomposite is able to respond to low intensity electric fields (applied voltage from 2.5 V to 5 V), with potential applications for the development of artificial smart micro-structures with motility behavior inspired by that of natural ciliate systems.

Keywords: polymer/metal nanocomposites; electroactive actuators; soft robotics

1. Introduction

Advances in nanotechnology and materials science are fueling new research paths in many engineering fields. A bursting trend is soft robotics, where the boundary between materials and devices is increasingly blurring in an attempt at mimicking biological systems [1]. New soft materials and microfabrication techniques enable devices that have been traditionally a prerogative of stiff components, such as actuators, sensors and even structural elements. In this sense, there is a need to create soft active matter, where materials can mechanically respond to external stimuli or vice versa. Many soft materials have been developed to deform in the presence of magnetic, fluidic, thermal, chemical and electrical stimuli [2]. Among those, electro-active polymers (EAPs) materials deform when subjected to an electric field. EAPs are normally divided in two sub-categories: dielectric and ionic materials, according to their deformation mechanism [3]. In ionic materials, the deformation is caused by a differential swelling of the polymer induced by the migration of ions in response to an applied electric field [4,5]. Ionic Polymer/Metal Composites (IPMCs) harness this mechanism when a voltage difference is applied between two electrodes embedded in the ionic polymer. Traditionally, the ionic polymer is Nafion or Flemion, whose surfaces are plated or coated with gold or platinum [6]. However, those materials have a relatively high Young modulus and traditional metallization techniques are not suitable for the production of stable, compliant and well-adherent electrodes [7,8].

Recently, we have developed a new type of IPMC material using ionogels and cluster-assembled nanostructured electrodes produced by Supersonic Cluster Beam Deposition (SCBD) to realize soft bending actuators [9,10]. This metallization approach relies on the use of supersonic beams seeded with neutral metal nanoparticles that are directed towards the ionogel target at room temperature. SCBD enables the high-throughput fabrication of thin conductive electrodes, highly resilient against deformation, without significantly altering the mechanical properties of the polymer [10]. However, those actuators are limited by their planar shape, whereas different morphologies would allow more complex deformation patterns and optimize the response according to the application. Soft artificial microswimmers, for example, inspired by the movement of biological cilia and flagella, have better hydrodynamic performances when designed with cylindrical shapes [11].

Here, we report on a new fabrication process of cylindrical ionogel/gold nanocomposites by combining replica molding and SCBD to produce 3D micro-actuators with high-aspect-ratio. The base material is a hydrophilic soft co-polymer that is able to incorporate large amounts of an imidazolium-based ionic liquid. After SCBD fabrication, the nanocomposite exhibited large electrochemical capacitance, suitable ionic conductivity and electromechanical response under the application of low intensity electrical stimuli (from 2.5 V to 5 V). The proposed manufacturing strategy has potential to be employed for the realization of smart electroactive structures for bio-inspired soft micro-components fabrication.

2. Materials and Methods

2.1. Ionogel Synthesis and Thin Films Fabrication

The ionogel matrix is a co-polymer based on 2-hydroxyethylmethacrylate (HEMA) and acrylonitrile (AN), chemically cross-linked using polyethylene glycole diacrylate (PEGDA, 500 Mw) under UV light irradiation. The polymer formulation was engineered to favor the incorporation of an imidazolium-based ionic liquid into the network, 1-(2-hydroxyethyl)-3-methylimidazolium tetrafluoroborate (OHMIM) [12]. This compound provides the presence of free ions in the material, which are able to migrate across the polymeric matrix under the application of an electric field. The polymer synthesis took place at room temperature in ambient conditions by dissolving the monomers and cross-linker into the ionic liquid. The ratio between the monomers was kept as HEMA: AN = 1.6 *w/w*, while the cross-linker amount was 0.02 *v/w* with respect to the total monomers' weight. The ratio between the monomers and ionic liquid was fixed at 1.24 *w/w*. After mixing the reagents into a 4 mL test tube (total mixture volume per sample was 1.5 mL), a 15 mg/µL solution of the photo-initiator (2,2-Dimethoxy-2-phenylacetophenone) in dimethyl sulfoxide solution was added into the pre-polymer (30 µL per stock solution). The relative amount of reagents was chosen on an empirical basis in order to obtain a soft elastic gel with elastomeric-like mechanical properties after UV photo-crosslinking at 365 nm. More specifically, the amount of monomers was calibrated in order to provide the gel with hydroxyl groups, present in the HEMA monomer, to favor the interaction with the OHMIM and to simultaneously toughen the matrix with the AN monomer. The balance between the ionic liquid phase and polymeric matrix confers the overall macroscopic softness and elasticity to the material. The latter was tuned and enhanced by the introduction of a high molecular weight cross-linker in the structure.

We used a simple molding approach to produce thin films of the prepared ionogel (namely, poly(HEMA-co-AN)/OHMIM). The fabrication consisted in pouring the pre-polymer solution into a spacer (0.5 mm thick) enclosed between two microscope glass slides and by exposing the system to UV light (light source was an UVP BlackRay, 100 W) for 60 min. These films served as benchmark reference samples to assess the electrical, electrochemical and electromechanical properties of the material after SCBD metallization.

2.2. Replica Molding

We employed replica molding to fabricate high-aspect ratio (>20) cylindrical ionogel-based micro-structures. This approach relied on the use of commercial syringe needles (0.5 mm diameter) as masters to generate a negative mold in polydimethylsiloxane (PDMS). PDMS is an elastomer with high chemical and thermal stability, widely employed for the molding of microfluidic devices and soft robotic components due to its ability to accurately replicate structures down to fractions of microns [13]. We chose metal needles as template because of their suitable size and shape, ease of demolding from the cured elastomer and for their easy purchase availability. The replica molding process to produce ionogel-based micro-components is depicted in Figure 1.

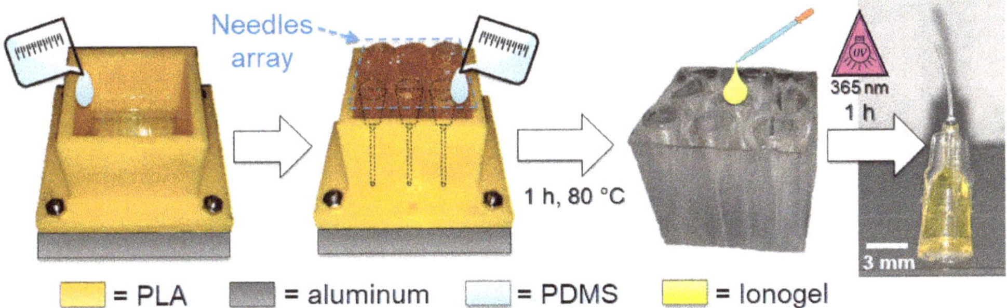

Figure 1. Schematic of the replica molding process to produce ionogel-based cylindrical micro-structures.

A 3 × 3 array of needles was positioned inside a dedicated box (28 mm × 28 mm wide, 30 mm high) fabricated by means of fused filament fabrication (FFF) 3D printing using poly lactic acid (PLA) and a Delta 2040 machine from WASP (Massa Lombarda (RA), Italy). The box was fastened to an aluminum basis that was covered with a first layer of PDMS (around 1 mm thick, elastomer: curing agent ratio = 10:1 w/w) where the needle tips were fixed. The box was then filled by casting PDMS (prepared with the same elastomer to curing agent ratio) in between the needles and up to the free surface. After polymerization in an oven at 80 °C for one hour, the new master mold was obtained by removing the needles from the polymer and unfastening the box. The negative replicas of the needles were then filled with the ionogel pre-polymer solution using a pipette and the mold was exposed to UV light at 365 nm for two hours. After complete cross-linking, cylindrical ionogels were demolded and dried in a vacuum overnight prior to further processing and testing.

2.3. Supersonic Cluster Beam Deposition for Ionogel/Gold Nanocomposites Fabrication

Supersonic Cluster Beam Deposition (SCBD) is a well-established metallization technique extensively employed for the fabrication of polymer/metal nanocomposites with controlled electrical and functional properties [10,14–16]. SCBD relies on the use of highly collimated supersonic beams seeded with neutral metal nanoparticles directed onto a target substrate to fabricate nanostructured metallic films with thickness in the range of hundreds of nm (Figure 2). The details on the operational principle are reported elsewhere [17] and summarized here below. A pulsed flux of inert gas is injected inside the cluster source at high pressure (40 bar), where a rotating metallic rod is subjected to high-voltage electric discharges (700 V). The atoms sputtered from the target metal then condense in neutral clusters. Due to a pressure gradient, the nanoparticles are carried by the gas expansion into a second chamber, forming a cluster-seeded supersonic beam [18]. The beam then passes through a skimmer and reaches the deposition chamber, where clusters impinge onto the target substrate. A quartz microbalance is used to monitor the amount of deposited material in real time. The thickness of the metal layer is solely dependent on the number of clusters deposited, which is in turn determined by the deposition time [17]. This process parameter simultaneously controls the resulting surface roughness of the film at the nanoscale, with typical values ranging from several nanometers to a few tens of nanometers [19,20].

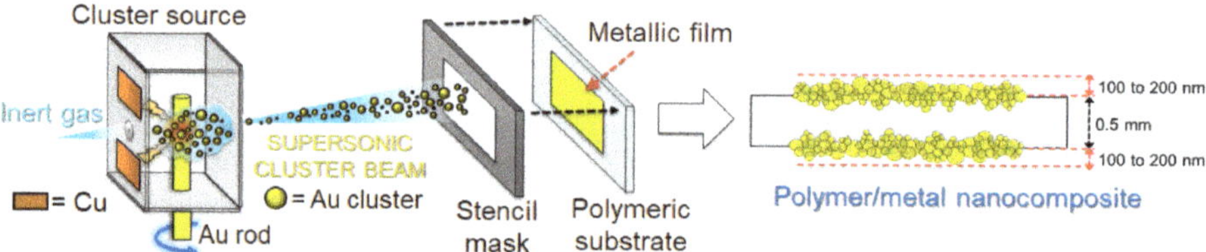

Figure 2. Simplified schematic of the SCDB process for polymer/metal nanocomposites fabrication.

Due to the low kinetic energy (around 0.5 ev/atom), they do not undergo significant deformation or fragmentation upon impact with the target [17]. When using soft polymers as deposition substrates, the metallic film partially interpenetrates with the polymeric network, resulting in the formation of a polymer/metal nanocomposite, with no damage or carbonization of the polymer [14]. When dealing with the metallization of ionic electroactive polymers, these features are particularly advantageous. In fact, the generation of a cluster-assembled metallic layer inside the polymeric matrix enhances the ionic charge accumulation efficiency under the application of an electric field at the electrodes, due to the available large surface area and porosity of the metal phase [9]. As compared to other classical metal deposition techniques, such as atomic layer deposition [21,22], SCBD enables a physical interlocking of the metallic structure to the polymer rather than the fabrication of a surface thin film. Moreover, it does not require surface functionalization of the substrate or the presence of precursor molecules and reactive surface groups on the target to favor the deposition process. It also avoids the triggering of chemical reactions into the polymers upon fabrication and takes place at room temperature. The SCBD process to fabricate polymer/metal nanocomposites is schematized in Figure 2.

We used this approach to fabricate conductive gold electrodes with large surface area on ionogel-based thin films and cylindrical micro-structures. SCBD was performed using stencil masks in both cases to define the geometry of the electrodes on the opposite sides of the samples, obtaining a classical capacitor-like geometry. The equivalent thickness of the fabricated layer (i.e., the thickness of a thin film produced on a rigid flat substrate using the same amount of clusters) was around 150 nm. The amount of gold deposited on the ionogel was measured in-situ using the dedicated quartz micro-balance. The equivalent thickness of the composite was assessed off-line using a profilometer (KLA Tencor, Milpitas, CA, USA), measuring the thickness of a cluster-assembled film deposited on a reference silicon substrate, mounted beside the target ionogel during deposition.

For planar ionogel films, the electrodes' surface area was defined by an 8 mm × 8 mm squared region using hollow aluminum masks (produced through milling). For the metallization of ionogel-based cylindrical structures, we designed and fabricated, by means of stereolithograpic 3D printing, a sample support able to fit the ionogels, as represented in Figure 3. The printer employed was a Form 2 from FormLabs, while the material was a standard FormLabs Clear Resin. With this approach, the electrodes fabricated on the cylindrical micro-structures were 0.4 mm wide metallic strips covering the whole length of the needle replica, avoiding electrical contact between the metallized regions of the nanocomposite. The ionogels were fully metallized using rastering, with the objective of connecting the larger basis underlying the cylinders (i.e., the ionogel replica of the plastic connector for syringe fitting) with external electronic equipment to prompt the actuation of the micro-structure. The sheet resistance of both nanocomposite cylinders and thin films was assessed using a multimeter.

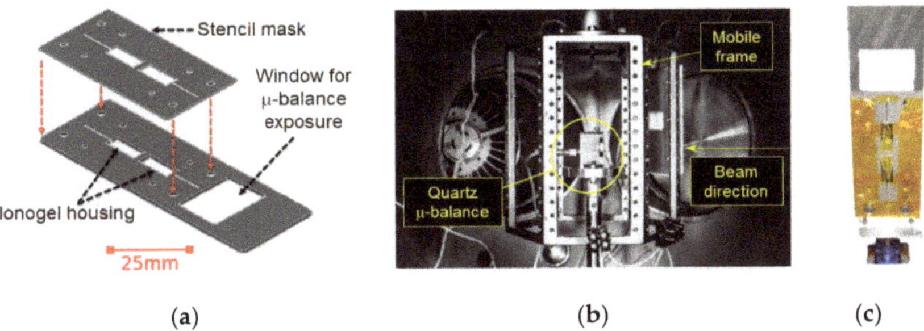

Figure 3. (**a**) Computer-Aided Design (CAD) drawing of the sample support used for the fabrication of cylindrical ionogel/metal nanocomposites; (**b**) photograph of the deposition chamber where samples are mounted on a mobile holder for metallization. The holder can rotate and translate in the zenithal direction during deposition using a programmable remote controller; (**c**) photograph of the assembled support system for cylindrical ionogel/metal nanocomposite fabrication.

2.4. Electrochemical and Electromechanical Characterization

The electrochemical properties of the nanocomposites were tested using electrochemical impedance spectroscopy (EIS) and cyclic voltammetry (CV) on the thin film samples. These techniques provide relevant information on the charge accumulation ability and resistive behavior of the nanocomposites, as well as on their electrochemical stability [23]. EIS was conducted by applying a sinusoidal voltage signal to the nanocomposite and by recording the corresponding current response. We kept the voltage amplitude at 5 mV, while the frequency was varied between 10^6 Hz and 10^{-2} Hz. The system impedance is determined by the phase shift between the input and output signals and current amplitude [16]. CVs were carried out by applying a voltage ramp varying between −5 V and 5 V, with a scan rate of 10 mV/s, and by recording the induced current. A typical voltammogram displays the presence of symmetric current peaks located at the voltage values where redox reactions take place at the electrodes. The equipment used for these tests was a Gamry Reference 600 potentiostat.

The actuation behavior of the samples was assessed using a dedicated setup consisting of a waveform generator and a power supply, connected to a custom-designed control circuit, to apply electrical sinusoidal signals of different frequencies (0.1 Hz to 10 Hz) and amplitudes (1 V to 5 V) to the nanocomposites. We used thin metal plates (0.2 mm thick) as interconnects, employing a custom-designed FFF printed support to regulate the clamping pressure of the electrical contact to the actuators. The plates were made of commercial copper-based layers coated in-house with a thin gold film (200 nm thick, produced by means of physical vapor deposition) to provide electrical connections with suitable electrical and electrochemical stability in the voltage range explored [12]. The linear displacement of the benders was registered in real time using a camera, and the acquired images were then processed by means of a MATLAB script developed in-house. This analysis tool permits tracking of the motion of the actuator tip along the deformation path using the Object Tracking algorithm of the MATLAB Computer Vision Toolbox.

3. Results and Discussions

3.1. Ionogel Molding and Nanocomposite Fabrication

The replication technique was very effective in reproducing the ionogel cylindric samples with good dimensional accuracy. Samples dimensions were observed and compared to the masters using transmission and reflection optical microscopy. The average diameter of the ionogel cylinders was 485 μm, whereas the needles' diameter was measured as 515 μm (both values calculated over ten measurements conducted along the cylinder) with a percentage standard deviation of roughly 1%. This size discrepancy between the metal master and ionogel diameter is fairly acceptable, considering the drying process that the polymers underwent. We analyzed the moisture-sensitivity character of the ionogel by measuring the weight of thin film samples straight after synthesis and every two hours under exposure to ambient conditions (RH = 45%, T = 23 °C). We then calculated the weight increase with respect to the initial weight, which was systematically found to be 3.5% (measurements were conducted on five samples). To the best of our knowledge, there is no report in the literature on fabrication approaches enabling the production of high-aspect-ratio cylindrical electroactive structures based on soft polymers at this length scale in 3D.

For ionogel/gold nanocomposites fabrication, we decided to use an equivalent thickness of the deposited metal equal to 150 nm in order to provide the ionogels with well-adherent conductive electrodes for efficient electrical signal transmission. In general, the increase in the electrical conductivity of a polymer during SCBD is observed in correspondence to a critical thickness of the forming metallic film [9,14,24]. At this transition point (the electrical percolation threshold), clusters start to form a sufficiently high number of physical connections to permit the flow of an electrical current in the system [24]. As the number of clusters increases, the film tends to behave like an ohmic conductor [14,24]. At this stage, a further increase in the layer thickness does not affect the electrical conductivity or the electrochemical performance of the composite material [9,14,24].

After SCBD metallization, the nanocomposites showed sheet resistance with values ranging from 50 to 100 Ohm*cm for the thin films and between 300 and 500 Ohm for the 10 mm long cylinder. Images of both thin films and cylindrical ionogel/gold nanocomposites are reported in Figure 4. This resistance increase can be attributed to the poor surface finishing of the metal masters, used as purchased without further polishing, which is reproduced on the ionogel surface after replica molding [25].

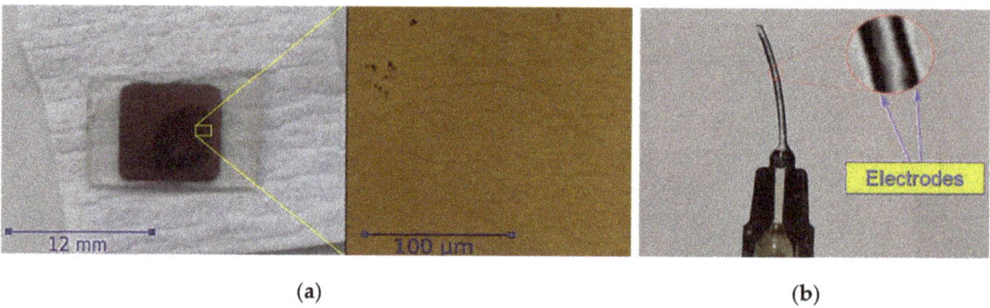

(a)   (b)

Figure 4. (a) Photograph and microscope images of an ionogel/gold nanocomposite film and (b) photograph of a nanocomposite cylindrical micro-actuator. Inset shows a magnification of the cylinder, highlighting the physical separation of the electrodes fabricated on the opposite sides of the structure.

We observed no electrical contact between the metal electrodes fabricated on the opposite sides of the cylinders (Figure 4b). This suggests that the stencil mask and sample support employed for SCBD metallization were suitable for the correct electrode patterning of the ionogel. However, the electrodes deposited on the cylindrical body were also shown to be electrically disconnected from the larger basis underlying the micro-actuator. This electrical discontinuity is probably due to the presence of abrupt cross-sectional changes along the ionogel replica in the direction parallel to the beam. These geometrical variations implied a physical gap in the deposited metal film located at the interface between the 0.5 mm diameter cylinder and the rest of the nanocomposite structure. This issue could be circumvented by employing custom-designed master molds with a smoother profile (e.g., by means of 3D printing VAT photopolymerization) for ionogel fabrication and by tilting the sample holder in the deposition chamber to favour the metallization of non-planar structures. This last solution has already proven to be successful for SCBD fabrication of 3D freeform polymer/metal nanocomposites [26,27].

SCBD is a high-throughput technique that can be employed for the scale-up manufacturing of soft actuators, as previously reported [9,10]. The substrate rastering in the direction perpendicular to the cluster beam can be exploited for the metallization of large areas [14], delivering gold clusters only on the substrate. It relies on the ablation of small metal precursors [17], with small amounts (around 1 mg/500 mm^2) and no significant waste of raw material.

3.2. Ionogel/Gold Nanocomposites Electrochemical and Actuation Properties

Results of the electrochemical analysis conducted on the nanocomposites are reported in Figure 5. In Figure 5a, it is possible to observe the Bode plot correlating the impedance modulus and phase with the frequency of the input voltage signal. A resistive behaviour was registered at intermediate frequencies where the phase value is close to 0°, while, in the low frequency domain, the presence of a double-layer capacitance was observed, with a maximum phase shift of about −60° at 10^{-2} Hz. By modelling the nanocomposite samples as simple RC circuits [9], we calculated the values of the specific capacitance and ionic conductivity, which was equal to 24 µF/cm^2 and 0.05 mS/cm, respectively.

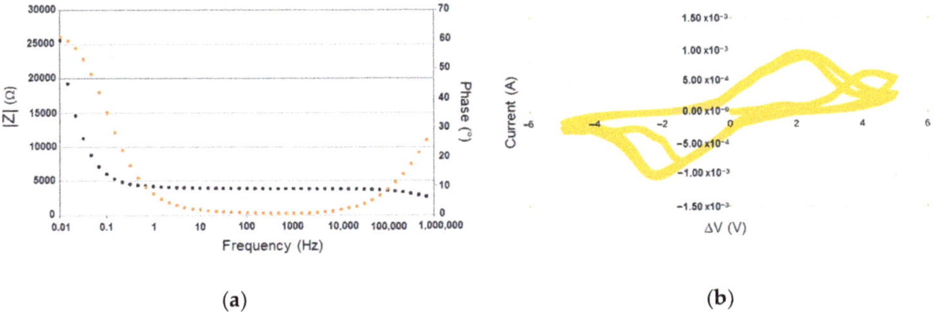

(a) **(b)**

Figure 5. Electrochemical analysis of the ionogel/gold nanocomposites: (a) Bode plot and (b) cyclic voltammograms registered with the potentiostat.

The value of the electrochemical capacitance is in line with that reported in the literature for this class of nanocomposites (several tens to a few hundreds of µF/cm^2 [10,16]), for which the large surface area of the cluster-assembled electrodes have a determinant role in favouring an efficient ionic charge accumulation. On the other hand, the ionic conductivity was moderately low when compared to other ionogel/metal nanocomposites produced by means of SCBD (typical values between 0.1 mS/cm and 1 mS/cm [10,16]), but still suitable to guarantee the migration of the charge carriers inside the polymeric matrix under the application of an electric field.

This behaviour can be associated with the lack of ionic groups (and relative solvated mobile counter-ions) in the polymer and with a significant interaction between the ionic liquid and the network backbone, damping the overall ionic mobility [28,29]. The voltammograms (Figure 5b) show the presence of symmetric current peaks in the mA order that can be ascribed to the activation of redox reactions. The electrochemical window of stability of the nanocomposites is between −2.5 V and 2.5 V.

Electromechanical actuation tests were first performed on 15 mm long and 1 mm wide nanocomposite strips (0.5 mm thick), manually cut from the fabricated thin films. The actuators underwent electrical stimulation in a cantilever configuration, clamped to a 3 mm × 3 mm wide support. Material reactivity to the application of electric fields started at 2.5 V. Before testing the composite frequency response, a 5 V continuous voltage was applied to the actuators to assess the maximum linear displacement of the tip, which was 0.3 mm. This limited motion can be ascribed to the combination of the ionogel softness and actuator thickness, the latter being higher with respect to the typical thickness (0.07 to 0.3 mm) of the ionic actuators described in the literature [5,7,30]. The percentage strain (ε) of the nanocomposites was calculated to have a more direct comparison with IPMCs actuators of different geometries [10]. More specifically, ε was computed as $\varepsilon = 2d \times t/(L^2 + d^2)$, where d is the measured linear tip displacement, t is the actuator thickness and L the cantilever free length. The nanocomposite ε value was 0.1%, which is compatible with the most recent results reported in the literature concerning soft ionic electroactive actuators [31,32].

Frequency tests were then carried out on the samples. The tip displacement tracking was conducted by applying sinusoidal voltage signals with a fixed amplitude of 3 V and 5 V. A graph showing the registered tip displacement over time at three different frequencies at 5 V is reported in Figure 6a. We could observe deformation of the actuators in phase with the applied electrical signal up to 10 Hz, with displacements decreasing as the frequency was increased. The actuation was shown to be stable and repeatable over time in the frequency range explored. Cylindrical nanocomposite micro-structures could not be tested in the same setup used for the cantilevers, due to the abovementioned electrical disconnection between the cylinder and the underlying interconnection basis and to criticalities encountered in the attempt to electrically connect the soft cylinder with the actuation control circuit. Therefore, we used remote electrodes, consisting of 10 cm × 10 cm wide stainless steel plates, to apply an electric field to cylindrical ionogel samples located between the electrodes (Figure 6b).

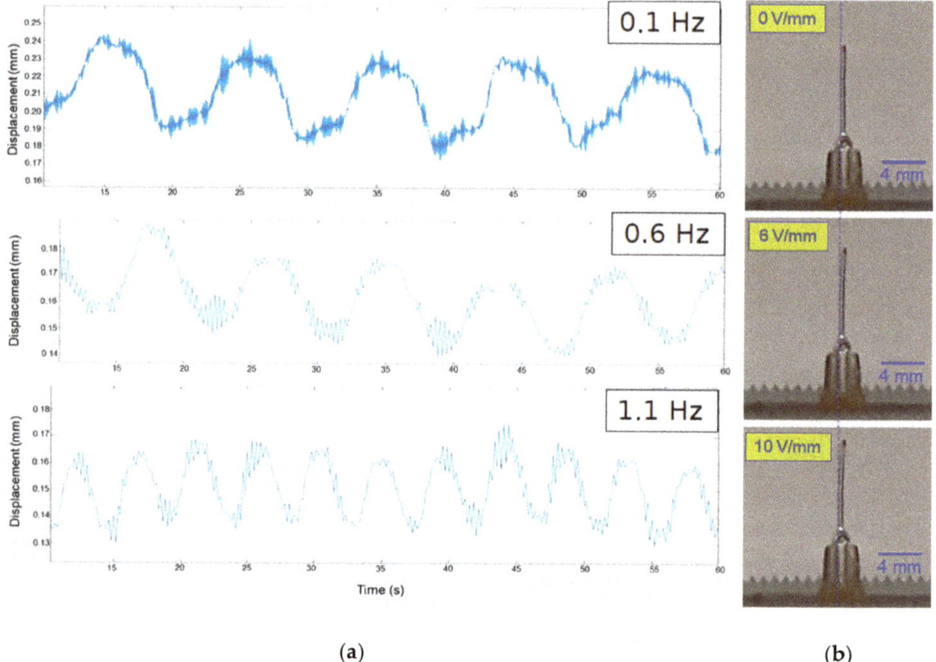

Figure 6. (**a**) Linear displacement of the cantilever nanocomposite actuator at 5 V for three frequency values. (**b**) Snapshots of the bending of an ionogel cylinder triggered using remote electrodes to apply the electrical stimulus.

The relative metal plate distance was 60 mm. To obtain electric field values comparable to that employed to trigger the actuation of the nanocomposite thin films (between 6 V/mm and 10 V/mm), we employed a high-voltage generator operating in the kV range. The obtained tip displacements for the cylinders were systematically lower, but comparable to that of the thin films (Figure 6b).

Although the charge accumulation mechanism in the case of remote electrodes differs from that taking place at metallic structures integrated into the actuators, these preliminary tests demonstrated the electroactivity of the high-aspect-ratio cylindrical ionogels. SCBD metallization of the micro-cylinders can be optimized by implementing several solutions. Masters with appropriate surface finishing and controlled roughness would favour the deposition and have potential to increase the electrodes' conductivity as well [14,33]. A more homogeneous electrical connection between the actuating body and basis interconnects of the ionogel could be obtained by identifying the critical tilting angles of the SCBD sample holder to effectively coat non-planar structures [27]. Moreover, the addition of suitable electrolytes and functional nanostructures into the ionogel matrix would be beneficial to further increase the material ionic conductivity and improve the actuation response in terms of displacement and response time [10].

The fabrication method proposed enables new applicative scenarios of soft devices with functional embodiments due to the synergy between replica moulding and SBCD. For example, cylindrical pillars can be used to form arrays of artificial electro-active cilia thanks to their slender geometry. Moreover, ionogel-based nanocomposites are not only interesting for their actuation characteristic, but also for their sensing ability. When an ionogel structure deforms, an electrical signal can be detected at the electrodes due to the variation of the ions concentrations inside the polymer matrix (piezo-ionic effect [12]). Therefore, this particular combination of geometry and materials can lead to a new generation of self-sensing actuators that open new horizons in small-scale soft robotics, from grippers with integrated haptic feedback to self-coordinated artificial cilia. The high electrochemical capacitance of the nanocomposites also has potential to be exploited for the design and manufacturing of energy storage functional devices with complex 3D architectures.

4. Conclusions

In this work, we report on the combination of replica molding and SCBD to fabricate high-aspect-ratio cylindrical structures of soft ionogel/gold nanocomposites. We used commercial metallic needles as masters to produce PDMS molds to fabricate ionogel-based cylinders (0.5 mm diameter, aspect ratio > 20). The cylinders are then metallized by means of SCBD to provide the polymer with integrated conductive gold nanostructure electrodes (150 nm thick). Nanocomposite thin films were used to assess the electrochemical and the electromechanical properties of the material, which revealed large electrochemical capacitance (24 $\mu F/cm^2$) for charge storage and a moderate ionic conductivity of about 0.05 mS/cm.

Actuation tests showed the electroactivity of the material in response to applied voltages ranging from 2.5 V to 5 V and a stable frequency response, while the actuation properties of the micro-cylinders were preliminarily investigated using remotely controlled electrodes. We foresee the implementation of such smart nanocomposite components for the realization of soft electroactive micro-systems with motion ability inspired by that of biological cilia.

Author Contributions: E.M. and T.S. conceived the research and wrote the manuscript; P.A. manufactured the structures and characterized the material; E.M., T.S. and L.M. critically analyzed the experimental results; P.M. supervised and supported the research. All authors revised the manuscript. All authors have read and agreed to the published version of the manuscript.

Funding: This research was supported by the Fund for Scientific Research-Flanders (FWO).

Conflicts of Interest: The authors declare no conflict of interest.

References

1. Rus, D.L.; Tolley, M.T. Design, fabrication and control of soft robots. *Nature* **2015**, *521*, 467–475. [CrossRef] [PubMed]
2. Hines, L.; Petersen, K.; Lum, G.Z.; Sitti, M. Soft Actuators for Small-Scale Robotics. *Adv. Mater.* **2016**, *29*. [CrossRef] [PubMed]
3. Carpi, F.; Kornbluh, R.; Sommer-Larsen, P.; Alici, G. Electroactive polymer actuators as artificial muscles: Are they ready for bioinspired applications? *Bioinspiration Biomim.* **2011**, *6*, 045006. [CrossRef] [PubMed]
4. Nardinocchi, P.; Pezzulla, M.; Placidi, L. Thermodynamically based multiphysic modeling of ionic polymer metal composites. *J. Intell. Mater. Syst. Struct.* **2011**, *22*, 1887–1897. [CrossRef]

5. Bhandari, B.; Lee, G.-Y.; Ahn, S.-H. A review on IPMC material as actuators and sensors: Fabrications, characteristics and applications. *Int. J. Precis. Eng. Manuf.* **2012**, *13*, 141–163. [CrossRef]

6. Wang, J.; Xu, C.; Taya, M.; Kuga, Y. A Flemion-based actuator with ionic liquid as solvent. *Smart Mater. Struct.* **2007**, *16*, S214–S219. [CrossRef]

7. Pugal, D.; Jung, K.; Aabloo, A.; Kim, K.J. Ionic polymer-metal composite mechanoelectrical transduction: Review and perspectives. *Polym. Int.* **2010**, *59*, 279–289. [CrossRef]

8. Okazaki, H.; Sawada, S.; Kimura, M.; Tanaka, H.; Matsumoto, T.; Ohtake, T.; Inoue, S. Soft Actuator Using Ionic Polymer–Metal Composite Composed of Gold Electrodes Deposited Using Vacuum Evaporation. *IEEE Electron Device Lett.* **2012**, *33*, 1087–1089. [CrossRef]

9. Yan, Y.; Santaniello, T.; Bettini, L.G.; Minnai, C.; Bellacicca, A.; Porotti, R.; Denti, I.; Faraone, G.; Merlini, M.; Lenardi, C.; et al. Electroactive Ionic Soft Actuators with Monolithically Integrated Gold Nanocomposite Electrodes. *Adv. Mater.* **2017**, *29*. [CrossRef]

10. Santaniello, T.; Migliorini, L.; Yan, Y.; Lenardi, C.; Milani, P. Supersonic cluster beam fabrication of metal–ionogel nanocomposites for soft robotics. *J. Nanoparticle Res.* **2018**, *20*, 250. [CrossRef]

11. Milana, E.; Gorissen, B.; Peerlinck, S.; De Volder, M.; Reynaerts, D. Artificial Soft Cilia with Asymmetric Beating Patterns for Biomimetic Low-Reynolds-Number Fluid Propulsion. *Adv. Funct. Mater.* **2019**, *29*, 1–8. [CrossRef]

12. Villa, S.M.; Mazzola, V.M.; Santaniello, T.; Locatelli, E.; Maturi, M.; Migliorini, L.; Monaco, I.; Lenardi, C.; Franchini, M.C.; Milani, P. Soft Piezoionic/Piezoelectric Nanocomposites Based on Ionogel/BaTiO3 Nanoparticles for Low Frequency and Directional Discriminative Pressure Sensing. *ACS Macro Lett.* **2019**, *8*, 414–420. [CrossRef]

13. Johnston, I.D.; McCluskey, D.K.; Tan, C.K.L.; Tracey, M.C. Mechanical characterization of bulk Sylgard 184 for microfluidics and microengineering. *J. Micromech. Microeng.* **2014**, *24*, 035017. [CrossRef]

14. Corbelli, G.; Ghisleri, C.; Marelli, M.; Milani, P.; Ravagnan, L. Highly Deformable Nanostructured Elastomeric Electrodes With Improving Conductivity Upon Cyclical Stretching. *Adv. Mater.* **2011**, *23*, 4504–4508. [CrossRef]

15. Santaniello, T.; Migliorini, L.; Borghi, F.; Yan, Y.; Rondinini, S.; Lenardi, C.; Milani, P. Spring-like electroactive actuators based on paper/ionogel/metal nanocomposites. *Smart Mater. Struct.* **2018**, *27*, 065004. [CrossRef]

16. Migliorini, L.; Santaniello, T.; Rondinini, S.; Saettone, P.; Franchini, M.C.; Lenardi, C.; Milani, P. Bioplastic electromechanical actuators based on biodegradable poly(3-hydroxybutyrate) and cluster-assembled gold electrodes. *Sens. Actuators B Chem.* **2019**, *286*, 230–236. [CrossRef]

17. Wegner, K.; Piseri, P.; Tafreshi, H.V.; Milani, P. Cluster beam deposition: A tool for nanoscale science and technology. *J. Phys. D Appl. Phys.* **2006**, *39*, R439–R459. [CrossRef]

18. Piseri, P.; Tafreshi, H.V.; Milani, P. Manipulation of nanoparticles in supersonic beams for the production of nanostructured materials. *Curr. Opin. Solid State Mater. Sci.* **2004**, *8*, 195–202. [CrossRef]

19. Borghi, F.; Podestà, A.; Piazzoni, C.; Milani, P. Growth Mechanism of Cluster-Assembled Surfaces: From Submonolayer to Thin-Film Regime. *Phys. Rev. Appl.* **2018**, *9*, 044016. [CrossRef]

20. Mirigliano, M.; Borghi, F.; Podestá, A.; Antidormi, A.; Colombo, L.; Milani, P. Non-ohmic behavior and resistive switching of Au cluster-assembled films beyond the percolation threshold. *Nanoscale Adv.* **2019**, *1*, 3119–3130. [CrossRef]

21. Johnson, R.W.; Hultqvist, A.; Bent, S.F. A brief review of atomic layer deposition: From fundamentals to applications. *Mater. Today* **2014**, *17*, 236–246. [CrossRef]

22. Guo, H.C.; Ye, E.; Li, Z.; Han, M.-Y.; Loh, X.J. Recent progress of atomic layer deposition on polymeric materials. *Mater. Sci. Eng. C* **2017**, *70*, 1182–1191. [CrossRef] [PubMed]

23. Kusoglu, A.; Weber, A.Z. Electrochemical/Mechanical Coupling in Ion-Conducting Soft Matter. *J. Phys. Chem. Lett.* **2015**, *6*, 4547–4552. [CrossRef] [PubMed]

24. Ravagnan, L.; Divitini, G.; Rebasti, S.; Marelli, M.; Piseri, P.; Milani, P. Poly(methyl methacrylate)–palladium clusters nanocomposite formation by supersonic cluster beam deposition: A method for microstructured metallization of polymer surfaces. *J. Phys. D Appl. Phys.* **2009**, *42*, 082002. [CrossRef]

25. Cardia, R.; Melis, C.; Colombo, L. Neutral-cluster implantation in polymers by computer experiments. *J. Appl. Phys.* **2013**, *113*, 224307. [CrossRef]

26. Bellacicca, A.; Santaniello, T.; Milani, P. Embedding electronics in 3D printed structures by combining fused filament fabrication and supersonic cluster beam deposition. *Addit. Manuf.* **2018**, *24*, 60–66. [CrossRef]

27. Santaniello, T.; Milani, P. Additive Nano-Manufacturing of 3D Printed Electronics Using Supersonic Cluster Beam Deposition. In *Frontiers of Nanoscience*; Elsevier: Amsterdam, The Netherlands, 2020; Volume 15, pp. 313–333.

28. Le Bideau, J.; Viau, L.; Vioux, A. Ionogels, ionic liquid based hybrid materials. *Chem. Soc. Rev.* **2011**, *40*, 907–925. [CrossRef]

29. Ding, Y.; Zhang, J.; Chang, L.; Zhang, X.; Liu, H.; Jiang, L. Preparation of High-Performance Ionogels with Excellent Transparency, Good Mechanical Strength, and High Conductivity. *Adv. Mater.* **2017**, *29*. [CrossRef]

30. Kim, K.J.; Tadokoro, S. Electroactive Polymers for Robotic Applications. *Artif. Muscles Sens.* **2007**, *23*, 291. [CrossRef]

31. Bian, C.; Zhu, Z.; Bai, W.; Chen, H.; Li, Y. Fast actuation properties of several typical IL-based ionic electro-active polymers under high impulse voltage. *Smart Mater. Struct.* **2020**, *29*, 035014. [CrossRef]

32. Aabloo, A.; Belikov, J.; Kaparin, V.; Kotta, U. Challenges and Perspectives in Control of Ionic Polymer-Metal Composite (IPMC) Actuators: A Survey. *IEEE Access* **2020**, *8*, 121059–121073. [CrossRef]

33. Marelli, M.; Divitini, G.; Collini, C.; Ravagnan, L.; Corbelli, G.; Ghisleri, C.; Gianfelice, A.; Lenardi, C.; Milani, P.; Lorenzelli, L. Flexible and biocompatible microelectrode arrays fabricated by supersonic cluster beam deposition on SU-8. *J. Micromech. Microeng.* **2011**, *21*, 45013. [CrossRef]

Publisher's Note: MDPI stays neutral with regard to jurisdictional claims in published maps and institutional affiliations.

 *applied nano*

Article

Effect of Ag Nanoparticle Size on Ion Formation in Nanoparticle Assisted LDI MS

Vadym Prysiazhnyi [1,2,*], **Filip Dycka** [2], **Jiri Kratochvil** [2], **Vitezslav Stranak** [2] **and Vladimir N. Popok** [3]

1   Faculty of Science, Masaryk University, Kamenice 5, 62500 Brno, Czech Republic
2   Faculty of Science, University of South Bohemia, Branisovska 1760, 37005 Ceske Budejovice, Czech Republic; fdycka@prf.jcu.cz (F.D.); jkratochvil@prf.jcu.cz (J.K.); stranak@prf.jcu.cz (V.S.)
3   Department of Materials and Production, Aalborg University, Skjernvej 4A, DK-9220 Aalborg, Denmark; vp@mp.aau.dk
*   Correspondence: prysiazhnyi@mail.muni.cz

Received: 8 July 2020; Accepted: 19 August 2020; Published: 24 August 2020

Abstract: Metal nanoparticles (NPs) were reported as an efficient matrix for detection of small molecules using laser desorption/ionization mass spectrometry. Their pronounced efficiency is mostly in desorption enhancement, while, in some cases, NPs can facilitate charge transfer to a molecule, which has been reported for alkali metals and silver. In this work, we present the study of the influence of Ag NP size on the laser desorption/ionization mass spectra of a model analyte, the molecule of riboflavin. The NPs were produced by magnetron sputtering-based gas aggregation in a vacuum and mass-filtered before the deposition on substrates. It was found that the utilization of smaller Ag NPs (below 15 nm in diameter) considerably enhanced the molecule desorption. In contrast, the laser irradiation of the samples with larger NPs led to the increased ablation of silver, resulting in *[analyte + Ag]*$^+$ adduct formation.

Keywords: silver nanoparticles; nano-PALDI MS; SALDI MS; laser desorption

---

1. Introduction

Surface enhancement employing nanomaterials in detection and imaging is implemented in many applications, including mass spectrometry [1], fluorescence imaging [2], Raman scattering [3–5], and various sensing technologies [6]. Often, the nanomaterials are prepared as nanoparticles (NPs), as there is a number of efficient production methods [7] also allowing the tailoring of the NP properties [8]. In particular, metal NPs are widely used in laser desorption/ionization mass spectrometry (LDI MS) [9]. Irradiating the samples with short laser pulses creates a plume of species (neutrals, ions, and molecules in excited states). Ionized species can be extracted and analyzed. For a few reasons, an additional component is required (the so-called matrix) that enhances the desorption and facilitates the charging of the molecules of interest [10]. Metal NPs can be such a matrix; two methods in the literature are the so-called surface-assisted LDI MS (SALDI MS) [11] and nanoparticle-assisted LDI MS (nano-PALDI MS) [12]. Two effects are typically highlighted when NPs are used as a matrix in LDI MS: local heating [13] (resulting in desorption enhancement) and high surface reactivity (which can lead to improved selectivity [14] and, thus, the increased sensitivity of the method [15]). Moreover, the matrix should provide conditions for efficient ionization [16]. Commonly, a protonation of an analyte is expected. However, a NP can also be a source of ions, which can form adducts with the molecules of interest [17].

An empirical approach was employed in the majority of studies, where nanomaterials were used as a matrix for LDI MS [18–20]. Although the reported high sensitivity and selectivity are remarkable, only a few works have studied the fundamental aspects of the processes leading to the enhancement.

In particular, the physics of laser–nanoparticle interaction in the presence of biological tissue has been studied relatively poorly, and this is easy to explain: the system is very complex. Though a nanoparticle matrix is considered to become "hot" under a laser impulse, leading to the more frequent fragmentation of an analyte, it also allows ionizing molecules that are not typically ionized using other matrix types [21,22]. The cationization of analytes was investigated, mostly focusing on alkali metals, as those form single charged ions [23,24]. In particular, a competition between different cationization agents on the riboflavin molecule was investigated [25]. In many cases [26–28], the charge transfer process was studied in metal salts or nanostructured surfaces. The conclusions of that research are not directly transferrable to NP-based matrices.

Silver is known as an element that efficiently forms adducts with different analytes [17]. This is related to the close proximity of the localized surface plasmon resonance (LSPR) wavelengths of Ag NPs to the typical laser wavelengths used in LDI MS devices and, thus, to an efficient energy transfer from the laser to the NPs [29]. For example, in our previous work, we showed that gas-aggregated Ag NPs have an adsorption maximum at 367 nm, while one of the commonly used laser wavelengths in MALDI devices is 355 nm [30]. We reported that use of large Ag NPs (32 ± 6 nm in diameter) led to a formation of adducts between small molecules and silver atoms. However, there were no studies on the NP size and its impact on ion formation and MALDI efficiency. Moreover, the application of small-size Ag NPs (below 10 nm) is intensively exploited for the detection of small molecules [30,31] and mass spectrometry imaging [32,33], demonstrating the enhancement of a signal $[M + H]^+$ ion in the most cases, where M denotes the molecule of interest. Even though a formation of $[M + Ag]^+$ ions was observed for certain molecules, the underlying physical processes were not investigated. The current study addresses two issues regarding the effect of NP size: i) the ion formation of a small-molecular-weight analyte; ii) the material release from NPs and adduct formation with the analyte molecule.

2. Materials and Methods

The substrate choice was confined by conditions for the time-of-flight (TOF) MS. It was necessary to have relatively small-sized, thin, conductive-from-both-sides, and flexible substrates, which would not superinduce additional ions in the mass spectra. The substrates were made of 0.1 mm-thick polyethylene terephthalate (PET) sheets. Plastic sheets were cut into 12 × 12 mm pieces, cleaned in isopropanol, and dried in air. The 250 nm-thick Cu film was deposited on each side in a custom-made vacuum system equipped with a 2" water-cooled magnetron with a Cu target. The film thickness was measured using an AlphaStep D-500 profilometer from KLA Tencor. The sputtering was carried out using Ar under 4 Pa of pressure and an applied current of 150 mA (DC power, 43 W).

Riboflavin ($C_{17}H_{20}N_4O_6$; molecular weight, 376.37 g/mol; p.a. grade) was used as a model molecule. The molecule contains a flavin group, representing importance for medical applications [34], and, as mentioned above, was used in similar studies. While a common approach to depositing an analyte onto the substrate is droplet crystallization, our preliminary tests showed that it results in the low reproducibility of MS in terms of both ion composition and absolute intensity. We relate this with the non-homogeneous analyte crystallization and coffee-ring effects [35]. Instead, a fine riboflavin powder was sprinkled over the substrate surface. We refer, by "fine powder", to microparticles that had a submillimeter diameter. By doing this, we were able to perform laser shots on the riboflavin surface, avoiding the necessity of dealing with crystallization issues. The substrates were placed into the vacuum chamber, where the NPs were deposited as described below, stabilizing the powder on the surface.

Silver NPs were produced using a commercial nanocluster source (NC200U from Oxford Applied Research) attached to the custom-built vacuum system, which, together with the source, was called MaSCA [36]. The particles were formed by the magnetron sputtering of a silver target (99.99% purity) and aggregation of the sputtered material in a vacuum. The NPs were collimated into a beam and mass-filtered prior to deposition using an electrostatic quadrupole mass selector, providing standard deviations of the particle diameters within 10% [37]. After the size selection, the NPs were deposited on the Cu/PET substrates in a soft-landing regime [38]. More details about the MaSCA and operation conditions can be found elsewhere [39]. For the current experiments, the NPs were filtered at voltages 200, 500, 700, 1100, and 1600 V. The chosen range of voltages defining the particle sizes (diameters) was based on the best sustainability for the deposition of mass-filtered NPs. The beam currents and deposition times were controlled in order to obtain the same surface coverage of NPs for every filtering voltage, $(3 \pm 1) \cdot 10^{11}$ cm^{-2}.

The deposited NPs were studied by atomic force microscopy (AFM) using a Ntegra Aura system (from NT-MDT). Commercial Si cantilevers with a tip curvature radius $\leq$ 10 nm were employed. It was found that the copper films on the PET substrates had a nanogranular structure with relatively high roughness (root mean square, RMS = 11.5 nm), as can be seen in Figure 1. Such surface morphology made it difficult to distinguish the deposited nanometer-scale particles. Therefore, the NPs were also deposited in parallel on smooth Si substrates (RMS < 0.5 nm), which were used for the size analysis.

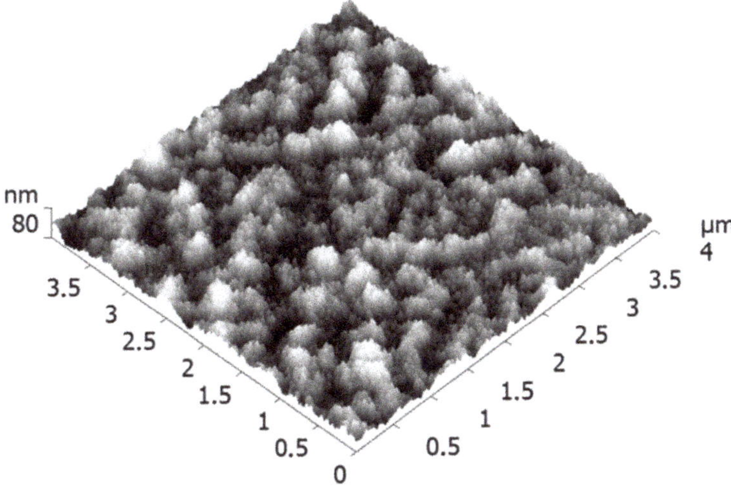

Figure 1. Atomic force microscopy (AFM) image of the copper film deposited on polyethylene terephthalate (PET).

For the LDI MS, the prepared samples were placed on ITO-covered glass using double-sided carbon tape and positioned on an MTP Slide Adapter II from Bruker. Mass spectra were acquired in the positive reflector ion mode using pulsed extraction with an extraction voltage of 16.3 kV, an acceleration voltage of 19 kV, a reflector voltage of 21 kV, a lens voltage of 8.3 kV, and a delayed extraction time of 130 ns on an autoflex™ speed mass spectrometer (from Bruker Daltonics). The laser was a Nd:YAG laser generating the third harmonic with a 355 nm wavelength. It produced pulses of 3 ns with an up to 2 kHz repetition rate. A single spectrum was acquired as an accumulation of 500 shots with a random walk pattern. Mass spectra were calibrated on silver ions (from 100 to 970 Da). Further processing was performed using the mMass software.

3. Results

3.1. Nanoparticle Characterization

An AFM image of a sample with NPs filtered at 200 V and deposited on the Si substrate is shown in Figure 2. The corresponding height histogram yields a mean value of 10.2 ± 1.0 nm. The height measurements were also carried for the particles deposited after filtering at 500, 700, 1100, and 1600 V (see Figures S1–S4 in Supplementary Materials). The particles filtered at adjacent voltages—for example,

500 and 700 or 700 and 1100 V—have a slight overlap in sizes due to the standard deviations. However, small voltage steps allowing the fine tuning (increase) of NP size can be essential for judging the role of NP diameter. These data are summarized in Table 1. The obtained values are in good agreement with the results previously published for NP size selection in MaSCA [34]. It is worth mentioning that silver NPs soft-landed in MaSCA typically retain their almost spherical shape; they only slightly oblate [40]. Thus, the AFM measurements of the particle heights can be a good estimate for the mean diameter/size. The particles filtered at 200, 500, 700, 1100, and 1600 V will be further referred to as 10, 15, 17, 20, and 23 nm NPs.

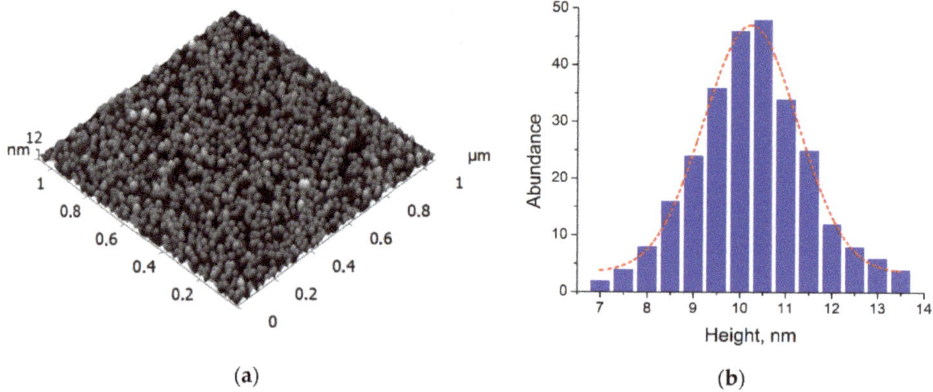

(a)   (b)

Figure 2. (**a**) AFM image and (**b**) height distribution of deposited Ag nanoparticles (NPs) filtered at 200 V.

Table 1. Mean values with standard deviations as obtained from AFM measurements for sizes of Ag NPs filtered at different voltages.

Filtering Voltage, V	200	500	700	1100	1600
NP height, nm	10.2 ± 1.0	14.6 ± 1.4	17.1 ± 1.6	19.5 ± 2.0	23.3 ± 2.4

3.2. Riboflavin Fragmentation

The riboflavin molecule consists of dimethyl isoalloxazine with an attached ribitol sugar. Figure 3 shows its molecular structure, highlighting possible bond cleavage spots. Two fragments with molecular weights of 242 (denoted as R1) and 135 g/mol (as R2) are observed in the mass spectra prepared with different matrices and under LDI mode, as the cleavage of the C–N bond comes up at lower laser fluence [41]. The cleavage of C–C bonds in ribitol sugar under higher laser fluence results in the appearance of charged fragments: adducts between the riboflavin molecule, its R1 part, cleaved parts of ribitol sugar (CH_2, CH_2COH_2, etc.), and Ag atoms. The mass spectra of riboflavin covered with Ag NPs have ions grouped into three categories. The first one is the silver clusters $Ag_n{}^+$ ($n \leq 9$). The intensities of Ag^+, $Ag_2{}^+$, and $Ag_3{}^+$ ions are found to be stable (within 10% error) for each measured mass spectrum, resulting in a ratio of 1:1.3:2.3, respectively. Silver clusters of larger size were detected irregularly, having low counts compared to those of Ag^+. The second category is the ions containing riboflavin molecules (*M*) (here and below, M refers to the riboflavin molecule), namely, $[M + H]^+$, $[M + Na]^+$, $[M + Ag]^+$, $[M + CH_2OH + H]^+$, and $[M + CH_2 + Ag]^+$. The third category of ions is the adducts formed with fragments of riboflavin (with H, Na, or Ag). Table 2 summarizes the experimental ion patterns and attributed molecular components. The observed ions were taken from a few randomly selected mass spectra for samples covered with NPs of different sizes, while it should be noted that the increase in laser energy and NP size led to the appearance of higher ion variety. The attribution is based on the mass/charge ratio (*m/z*), possible chemical structure, and isotopic pattern (modeled in mMass based on the chemical formula). The difference between the experimental and theoretical values is related to the spectrometer resolution. The typical full width at half maximum (FWHM) of the detected peaks is found between 0.1 and 0.4 Da, depending on the counts for particular ion. The intensity of the given mass peaks depends on the laser fluence and NP sizes. These effects are discussed below.

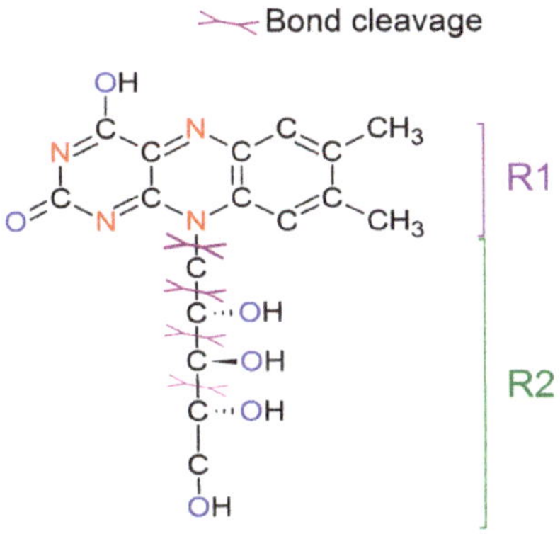

Figure 3. Structure of the riboflavin molecule with scheme of its fragmentation under laser irradiation.

Table 2. Attributed ions observed in the mass spectra of riboflavin covered by Ag NPs.

Attributed Structure	Experimental m/z	Theoretical m/z	Formula
$R2 + (CH_2) + Na$	172.07	172.07	$C_6H_{13}O_4Na$
$R2 + (CH_2)_2 + H$	186.06	186.08	$C_7H_{15}O_4Na$
$Ag_2 + H$	216.80	216.81	Ag_2H
$R2 + Ag$	241.81	241.97	$C_5H_{11}O_4Ag$
$R1 + H$	243.07	243.08	$C_{12}H_{11}N_4O_2$
$R2 + CH_2\text{-}H + Ag$	254.91	254.97	$C_6H_{12}O_4Ag$
$R1 + CH_2 + H$	257.09	257.10	$C_{13}H_{13}N_4O_2$
$R1 + Ag$	348.96	348.98	$C_{12}H_{10}N_4O_2Ag$
$R1 + CH_2 + Ag$	362.97	363.00	$C_{13}H_{12}N_4O_2Ag$
$M + H$	377.07	377.15	$C_{17}H_{21}N_4O_6$
$M + Na$	399.07	399.12	$C_{17}H_{20}N_4O_6Na$
$R1 + C_4H_{10}O_2 + Ag$	439.00	439.05	$C_{16}H_{20}N_4O_4Ag$
$R1 + Ag_2$	457.83	457.89	$C_{12}H_{10}N_4O_2Ag_2$
$M + Ag$	482.94	483.04	$C_{17}H_{20}N_4O_6Ag$
$M + CH_2\text{-}H + Ag$	497.03	497.05	$C_{18}H_{22}N_4O_6Ag$

3.3. Effect of Laser Fluence and NP Size on Mass Spectra of Riboflavin

The effect of laser fluence on the obtained mass spectra is shown in Figure 4. Note that laser fluence is calculated as a percentage of the total laser power (TLP), the only parameter provided by the TOF MS manufacturer. The lower fluence limit is determined by the threshold above which the number of generated ions start prevailing over the instrumental noise (40% of the TLP in our case) [42]. The upper fluence limit is related to the phenomenon of electrostatic repulsion in the ion cloud, causing m/z shifts, an increase in the FWHM, and signal distortion [43]. In our case, these effects become drastic when the fluence approaches 55% of the TLP.

At the lowest used laser fluence, the mass spectrum contains only protonated riboflavin, its protonated fragments, and $[M + Ag]^+$ ions of relatively low intensity. The absence of Ag_n^+ clusters and their adducts means that the transferred energy is not sufficient to eject Ag atoms from the NPs. The abovementioned clusters, as well as silver ions, and adducts with riboflavin/fragments start appearing at higher fluences.

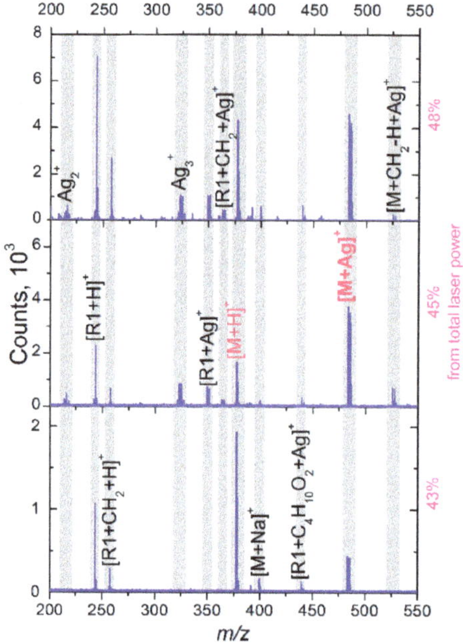

Figure 4. The effect of laser fluence on the mass spectra of riboflavin for the case of 10 nm Ag NPs.

Figure 5 shows the differences in the mass spectra for riboflavin covered with 10, 15, 17, and 23 nm Ag NPs (all the graphs have the same *y*-scale). In this case, the sample was irradiated with a fluence equivalent to 43% of the TLP. In the obtained spectra, one can see two main tendencies: (i) the intensity of protonated ions (for example, *[R1 + H]$^+$* and *[M + H]$^+$*) changes insignificantly with an increase in the particle size, while (ii) the ion counts related to silver clusters and adducts increase drastically for the cases of larger NPs. Two effects can be attributed to the observed phenomena: an increased interaction cross-section for the NPs of larger diameter (and, as a result, higher laser energy absorption) [44], and enhanced optical extinction due to the LSPR of Ag NPs [38,45]. Thus, the laser power transferred to an NP causes heating and a strong local electromagnetic field; both effects can facilitate the ejection of atoms or even small clusters. According to the Mie theory, the absorption cross-section depends on the cube of the particle radius [46]. Thus, one can suppose that for the smallest NPs, the absorbed energy is lower than a threshold value required for the ejection, while with the particle size increase, an ejection of ions and small clusters becomes possible, as can be seen in Figure 5. Since the thermal and electromagnetic phenomena interact for a metal NP in a sophisticated manner, the abovementioned hypothesis requires further study going beyond the goals of the current paper.

The integration of the measured signal obtained from the full isotopic pattern of a single molecule provides a value proportional to the number of generated ions. Based on the relative simplicity of our system (monomolecular sample and size-selected NPs), we can collect mass spectra of similar intensity from different samples and, thus, analyze the changes in ion counts through the absolute values of areas for the isotopic patterns. After initial data filtering (the removal of artifacts and mass spectra with low total ion counts), it is possible to compare the most prominent attributed structures depending on both the laser fluence and silver NP size. The results of the data evaluation are presented in Figure 6. The first important conclusion is that the absolute intensity of protonated riboflavin is independent (within errors) of the Ag NP size, while it increases with the laser fluence. We observe a slight decrease in *[M + H]$^+$* ion counts only for the highest laser fluences, which can be related to the significant increase in Ag ion intensity and consequent suppression effects [47]. The signal of *Ag$^+$* shows the same tendency to increase with particle size for all laser fluences, although the intensity values are higher for the higher fluences, which is well understood; higher power transfer leads to a higher level of material ablation. An important finding is that the NPs should be above a certain "threshold size" (around 15 nm in our case) to promote the materials' release, which is already discussed above (see Figure 5). This "threshold size" is also evident for the formation of *[M + Ag]⁺* ions. Within the errors of the experiments, the intensity of this signal does not considerably depend on either NP size or laser fluence.

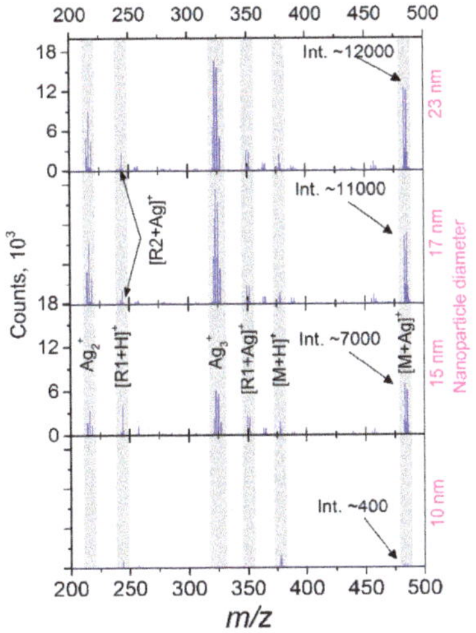

Figure 5. The effect of NP size on mass spectra of riboflavin for laser fluence equivalent to 43% of total laser power.

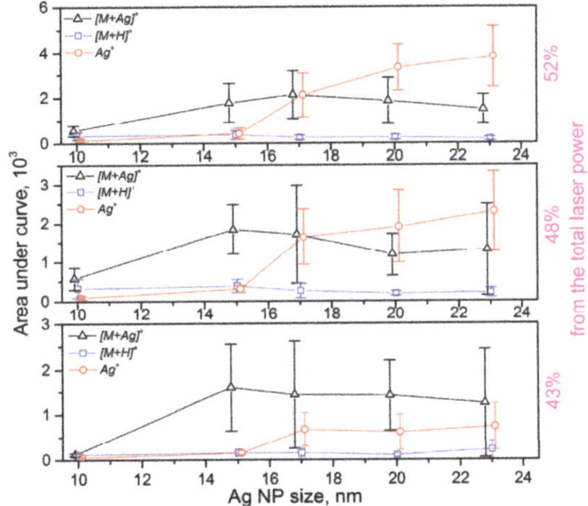

Figure 6. Comparison of absolute ion intensity for the selected ions as a function of Ag NP size for 43% and 48% of total laser power (TLP). Data points corresponding to the given NP sizes are intentionally slightly shifted on the *x*-axis for better visualization.

Figure 7 shows the ratio between the intensities for *[M + Ag]⁺* and *Ag⁺* ions for different laser fluences and NP sizes. It can be seen that for the lowest laser fluence and smallest NPs, the energy transfer is not enough to cause any considerable release of *[M + Ag]⁺* ions. For the higher fluences, the observed tendencies are the same concerning the particle size. The release of the silver ions is higher for the higher values of the laser fluence, and it has the same trend for the larger NPs. As for the ablated silver ions, it is evident that the utilization of NPs as a matrix would require the proper optimization of the material quantity (NP surface coverage) together with the laser fluence. For the proper measurement, the intensity of the silver ions should not significantly exceed the $[M + Ag]^+$ intensity. For example, for the TLP of 43%, the silver ion counts do not exceed those of the riboflavin/silver ions (see Figure 6). Meanwhile, for the TLP of 52%, the quantity of silver ions released from 17 nm NPs is equal to that of $[M + Ag]^+$.

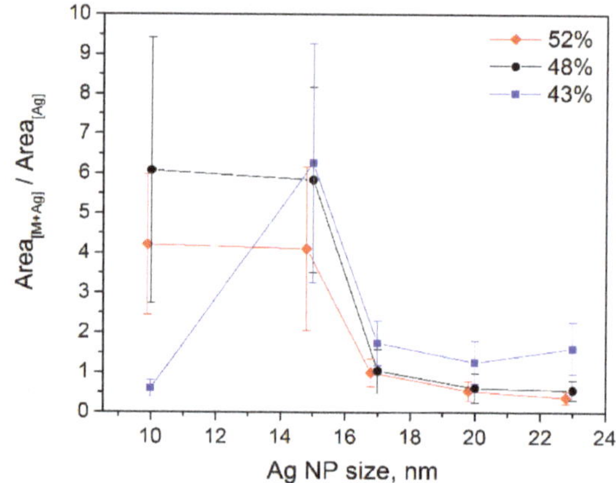

Figure 7. The ratio between $[M + Ag]^+$ and Ag^+ ion intensities depending on the Ag NP size and laser fluence.

4. Conclusions

Size-selected Ag NPs were prepared using the gas aggregation of magnetron-sputtered Ag and deposited on the substrates with the model analyte, riboflavin, in order to study the efficiency of NPs as a matrix for enhancing nano-PALDI MS. It has been shown that the release of Ag atoms, clusters, and ions from the NP surface and the formation of Ag adducts with the analyte molecule strongly depend on the NP size and laser fluence. Our study revealed that the silver release under the laser irradiation is insignificant for small NPs. However, it considerably increases for the particles with sizes ≥ 15 nm: silver ions and small charged silver clusters become the dominant species in the spectra. It is suggested that the particle heating and formation of a strong electromagnetic field (due to LSPR) upon interaction with laser light could be the main factors affecting the materials' release. The presence of metal ions in the plume triggers the formation of adducts with riboflavin and its fragments. In order to use the adduct ions for molecular detection, a proper amount of NP material in the matrix (requiring the adjustment of the NP size and surface coverage) should be used. One can also conclude that the use of matrices with large silver NPs could allow the investigation of other cationization agents.

Supplementary Materials: The following are available online at http://www.mdpi.com/2673-3501/1/1/2/s1. Figure S1: (**a**) AFM image and (**b**) height distribution of deposited Ag NPs filtered at 500 V; Figure S2: (**a**) AFM image and (**b**) height distribution of deposited Ag NPs filtered at 700 V; Figure S3: (**a**) AFM image and (**b**) height distribution of deposited Ag NPs filtered at 1100 V; Figure S4: (**a**) AFM image and (**b**) height distribution of deposited Ag NPs filtered at 1600 V.

Author Contributions: Conceptualization: V.P. and J.K.; data acquisition: V.P., F.D., and V.N.P.; data processing, V.P., F.D., J.K., and V.N.P.; original draft preparation, V.P.; review and editing, V.N.P., F.D., J.K., and V.S.; supervision, V.S.; All authors have read and agreed to the published version of the manuscript.

Funding: V.P. acknowledges the financial support from the Czech Science Foundation through the project GACR 19-20927Y and financial support from the Standard Trainership Erasmus+ program for staff mobility. F.D. was supported from the European Regional Development Fund-Project "Mechanisms and dynamics of macromolecular complexes: from single molecules to cells" (No. CZ.02.1.01/0.0/0.0/15_003/0000441). J.K. was supported from

TAČR GAMA-2 TP01010019. V.S. acknowledges the financial support from the Czech Science Foundation through the project GACR 19-20168S.

Acknowledgments: The authors acknowledge Antonin Bednarik for fruitful discussions on the MS data analysis.

Conflicts of Interest: The authors declare no conflict of interest. The funders had no role in the design of the study; in the collection, analyses, or interpretation of data; in the writing of the manuscript; or in the decision to publish the results.

References

1. Lei, C.; Qian, K.; Noonan, O.; Nouwens, A.; Yu, C. Applications of nanomaterials in mass spectrometry analysis. *Nanoscale* **2013**, *5*, 12033–12042. [CrossRef]
2. Keshavarz, M.; Tan, B.; Venkatakrishnan, K. Multiplex photoluminescent silicon nanoprobe for diagnostic bioimaging and intracellular analysis. *Adv. Sci.* **2018**, *5*, 1700548. [CrossRef]
3. Camden, J.P.; Dieringer, J.A.; Wang, Y.; Masiello, D.J.; Marks, L.D.; Schatz, G.C.; van Duyne, R.P. Probing the structure of single-molecule surface-enhanced Raman scattering hot spots. *J. Am. Chem. Soc.* **2008**, *130*, 12616–12617. [CrossRef] [PubMed]
4. Keshavarz, M.; Kassanos, P.; Tan, B.; Venkatakrishnan, K. Metal-oxide surface-enhanced Raman biosensor template towards point-of-care EGFR detection and cancer diagnostics. *Nanoscale Horiz.* **2020**, *130*, 294–307. [CrossRef]
5. Keshavarz, M.; Tan, B.; Venkatakrishnan, K. Label-free sers quantum semiconductor probe for molecular-level and in vitro cellular detection: A noble-metal-free methodology. *ACS Appl. Mater. Interfaces* **2018**, *10*, 34886–34904. [CrossRef] [PubMed]
6. Zeng, S.; Baillargeat, D.; Ho, H.P.; Yong, K.T. Nanomaterials enhanced surface plasmon resonance for biological and chemical sensing applications. *Chem. Soc. Rev.* **2014**, *43*, 3426–3452. [CrossRef]
7. Tsuzuki, T. Commercial scale production of inorganic nanoparticles. *Int. J. Nanotechnol.* **2009**, *6*, 567–588. [CrossRef]
8. Liz-Marzán, L.M. Tailoring surface plasmons through the morphology and assembly of metal nanoparticles. *Langmuir* **2006**, *22*, 32–41. [CrossRef]
9. Abdelhamid, H.N. Nanoparticle assisted laser desorption/ionization mass spectrometry for small molecule analytes. *Microchim. Acta* **2018**, *185*, 200. [CrossRef]
10. Gemperline, E.; Rawson, S.; Li, L. Optimization and comparison of multiple MALDI matrix application methods for small molecule mass spectrometric imaging. *Anal. Chem.* **2014**, *86*, 10030–10035. [CrossRef]
11. Tanaka, K.; Waki, H.; Ido, Y.; Akita, S.; Yoshida, Y.; Yoshida, T. Protein and polymer analyses up to m/z 100 000 by laser ionization time-of-flight mass spectrometry. *Rapid Commun. Mass Spectrom.* **1988**, *2*, 151–153. [CrossRef]
12. Taira, S.; Sugiura, Y.; Moritake, S.; Shimma, S.; Ichiyanagi, Y.; Setou, M. Nanoparticle-assisted laser desorption/ionization based mass imaging with cellular resolution. *Anal. Chem.* **2008**, *80*, 4761–4766. [CrossRef] [PubMed]
13. Boerigter, C.; Aslam, U.; Linic, S. Mechanism of Charge Transfer from Plasmonic Nanostructures to Chemically Attached Materials. *ACS Nano* **2016**, *10*, 6108–6115. [CrossRef] [PubMed]
14. Kailasa, S.K.; Wu, H.F. One-pot synthesis of dopamine dithiocarbamate functionalized gold nanoparticles for quantitative analysis of small molecules and phosphopeptides in SALDI- and MALDI-MS. *Analyst* **2012**, *137*, 1629–1638. [CrossRef]
15. Iwaki, Y.; Kawasaki, H.; Arakawa, R. Human serum albumin-modified Fe_3O_4 magnetic nanoparticles for affinity-SALDI-MS of small-molecule drugs in biological liquids. *Anal. Sci.* **2012**, *28*, 893–900. [CrossRef] [PubMed]
16. Zenobi, R.; Knochenmuss, R. Ion formation in MALDI mass spectrometry. *Mass Spectrom. Rev.* **2002**, *17*, 337–366. [CrossRef]
17. Lai, E.P.C.; Owega, S.; Kulczycki, R. Time-of-flight mass spectrometry of bioorganic molecules by laser ablation of silver thin film substrates and particles. *J. Mass Spectrom.* **1998**, *33*, 554–564. [CrossRef]
18. Teng, C.H.; Ho, K.C.; Lin, Y.S.; Chen, Y.C. Gold nanoparticles as selective and concentrating probes for samples in MALDI MS analysis. *Anal. Chem.* **2004**, *76*, 4337–4342. [CrossRef]

19. Dong, X.; Cheng, J.; Li, J.; Wang, Y. Graphene as a novel matrix for the analysis of small molecules by MALDI-TOF MS. *Anal. Chem.* **2010**, *82*, 6208–6214. [CrossRef]
20. Jiang, D.; Song, N.; Li, X.; Ma, J.; Jia, Q. Highly selective enrichment of phosphopeptides by on-chip indium oxide functionalized magnetic nanoparticles coupled with MALDI-TOF MS. *Proteomics* **2017**, *17*, 17–18. [CrossRef]
21. Lu, M.; Yang, X.; Yang, Y.; Qin, P.; Wu, X.; Cai, Z. Nanomaterials as Assisted Matrix of Laser Desorption/Ionization Time-of-Flight Mass Spectrometry for the Analysis of Small Molecules. *Nanomaterials* **2017**, *7*, 87. [CrossRef]
22. Jackson, S.N.; Baldwin, K.; Muller, L.; Womack, V.M.; Schultz, J.A.; Balaban, C.; Woods, A.S. Imaging of lipids in rat heart by MALDI-MS with silver nanoparticles. *Anal. Bioanal. Chem.* **2014**, *406*, 1377–1386. [CrossRef]
23. Goolsby, B.J.; Brodbelt, J.S.; Adou, E.; Blanda, M. Determination of alkali metal ion binding selectivities of calixarenes by matrix-assisted laser desorption ionization and electrospray ionization in a quadrupole ion trap. *Int. J. Mass Spectrom.* **1999**, *193*, 197–204. [CrossRef]
24. Hortal, A.R.; Hurtado, P.; Martínez-Haya, B.A.; Arregui, L. Bañares, Solvent-free MALDI investigation of the cationization of linear polyethers with alkali metals. *J. Phys. Chem. B* **2008**, *112*, 8530–8535. [CrossRef]
25. Yang, S.; Shi, Y.; Ma, Y.; Mu, L.; Kong, X. Competition between metal cationization and protonation/reduction in MALDI process: An example of riboflavin. *Int. J. Mass Spectrom.* **2018**, *434*, 209–214. [CrossRef]
26. Knochenmuss, R.; Dubois, F.; Dale, M.J.; Zenobi, R. The matrix suppression effect and ionization mechanisms in matrix-assisted laser desorption/ionization. *Rapid Commun. Mass Spectrom.* **1996**, *10*, 871–877. [CrossRef]
27. Wong, C.K.L.; Dominic Chan, T.-W. Cationization processes in matrix-assisted laser desorption/ionization mass spectrometry: Attachment of divalent and trivalent metal ions. *Rapid Commun. Mass Spectrom.* **1997**, *11*, 513–519. [CrossRef]
28. Montaudo, G.; Montaudo, M.S.; Puglisi, C.; Samperi, F.; Sepulchre, M. End-group characterization of poly(methylphenylsilane) by alkali metal salts doped MALDI-TOF mass spectra. *Macromol. Chem. Phys.* **1996**, *197*, 2615–2625. [CrossRef]
29. Wu, S.; Qian, L.; Huang, L.; Sun, X.; Su, H.; Gurav, D.D.; Jiang, M.; Cai, W.; Qian, K. A Plasmonic mass spectrometry approach for detection of small nutrients and toxins. *Nano-Micro Lett.* **2018**, *10*, 52. [CrossRef]
30. Prysiazhnyi, V.; Dycka, F.; Kratochvil, J.; Sterba, J.; Stranak, V. Gas-aggregated Ag nanoparticles for detection of small molecules using LDI MS. *Anal. Bioanal. Chem.* **2020**, *412*, 1037–1047. [CrossRef]
31. Sekuła, J.; Nizioł, J.; Rode, W.; Ruman, T. Silver nanostructures in laser desorption/ionization mass spectrometry and mass spectrometry imaging. *Analyst* **2015**, *140*, 6195–6209. [CrossRef] [PubMed]
32. Muller, L.; Baldwin, K.; Barbacci, D.C.; Jackson, S.N.; Roux, A.; Balaban, C.D.; Brinson, B.E.; McCully, M.I.; Lewis, E.K.; Schultz, J.A.; et al. Laser Desorption/Ionization Mass Spectrometric Imaging of Endogenous Lipids from Rat Brain Tissue Implanted with Silver Nanoparticles. *J. Am. Soc. Mass Spectrom.* **2017**, *28*, 1716–1728. [CrossRef] [PubMed]
33. Zhang, Z.; Li, S.; Liu, J.; Liu, L.; Yang, H.; Zhang, Y.; Wang, T.; Zhao, Z. Silver nanoparticles as matrix for MALDI FTICR MS profiling and imaging of diverse lipids in brain. *Talanta* **2018**, *179*, 624–631. [CrossRef]
34. Beztsinna, N.; Tsvetkova, Y.; Jose, J.; Rhourri-Frih, B.; al Rawashdeh, W.; Lammers, T.; Kiessling, F.; Bestel, I. Photoacoustic imaging of tumor targeting with riboflavin-functionalized theranostic nanocarriers. *Int. J. Nanomed.* **2017**, *12*, 3813–3825. [CrossRef] [PubMed]
35. Hu, J.B.; Chen, Y.C.; Urban, P.L. Coffee-ring effects in laser desorption/ionization mass spectrometry. *Anal. Chim. Acta.* **2013**, *766*, 77–82. [CrossRef]
36. Hanif, M.; Popok, V.N. Magnetron Sputtering Cluster Apparatus for Formation and Deposition of Size-Selected Metal Nanoparticles. In *Physics, Chemistry and Applications of Nanostructures*; Borisenko, V.E., Gaponenko, S.V., Gurin, V.S., Kam, C.H., Eds.; World Sci. Publ.: Singapore, 2015; pp. 416–419.
37. Hanif, M.; Juluri, R.R.; Chirumamilla, M.; Popok, V.N. Poly(methyl methacrylate) composites with size-selected silver nanoparticles fabricated using cluster beam technique. *J. Polym. Sci. Part. B Polym. Phys.* **2016**, *54*, 1152–1159. [CrossRef]
38. Popok, V. Energetic Cluster-Surface Collisions. In *Handbook of Nanophysics: Clusters and Fullerenes*; Sattler, K.D., Ed.; CRC Press: Boca Raton, CA, USA, 2010; pp. 19-1–19-19.
39. Popok, V.N.; Gurevich, L. Charge states of size-selected silver nanoparticles produced by magnetron sputtering. *J. Nanoparticle Res.* **2019**, *21*, 171. [CrossRef]

40. Novikov, S.M.; Popok, V.N.; Evlyukhin, A.B.; Hanif, M.; Morgen, P.; Fiutowski, J.; Beermann, J.; Rubahn, H.-G.; Bozhevolnyi, S.I. Highly-stable monocrystalline silver clusters for plasmonic applications. *Langmuir* **2017**, *33*, 6062–6070. [CrossRef]

41. McMahon, J.M. The analyses of six common vitamins by laser desorption mass spectroscopy. *Anal. Biochem.* **1985**, *147*, 535–545. [CrossRef]

42. Spengler, B.; Bahr, U.; Karas, M.; Hillenkamp, F. Postionization of laser-desorbed organic and inorganic compounds in a time of flight mass spectrometer. *Instrum. Sci. Technol.* **1988**, *17*, 173–193. [CrossRef]

43. Ingendoh, A.; Karas, M.; Hillenkamp, F.; Giessmann, U. Factors affecting the resolution in matrix-assisted laser desorption-ionization mass spectrometry. *Int. J. Mass Spectrom. Ion Process.* **1994**, *131*, 345–354. [CrossRef]

44. Govorov, A.O.; Richardson, H.H. Generating heat with metal nanoparticles. *Nano Today* **2007**, *2*, 30–38. [CrossRef]

45. González, A.L.; Noguez, C.; Beránek, J.; Barnard, A.S. Size, shape, stability, and color of plasmonic silver nanoparticles. *J. Phys. Chem. C* **2014**, *118*, 9128–9136. [CrossRef]

46. Hlaing, M.; Gebear-Eigzabher, B.; Roa, A.; Marcano, A.; Radu, D.; Lai, C.Y. Absorption and scattering cross-section extinction values of silver nanoparticles. *Opt. Mater.* **2016**, *58*, 439–444. [CrossRef]

47. Chandler, J.; Haslam, C.; Hardy, N.; Leveridge, M.; Marshall, P. A Systematic Investigation of the Best Buffers for Use in Screening by MALDI–Mass Spectrometry. *SLAS Discov.* **2017**, *22*, 1262–1269. [CrossRef]

applied nano

Article

Influence of the Parameters of Cluster Ions on the Formation of Nanostructures on the KTP Surface

Ivan V. Nikolaev * and Nikolay G. Korobeishchikov

Department of Applied Physics, Novosibirsk State University, Pirogov Str. 2, 630090 Novosibirsk, Russia; korobei@ci.nsu.ru
* Correspondence: i.nikolaev@nsu.ru

Abstract: In this work, the formation of periodic nanostructures on the surface of potassium titanyl phosphate (KTP) has been demonstrated. The surface of KTP single crystals after the processing of argon cluster ions with different energy per cluster atom E/N_{mean} = 12.5 and 110 eV/atom has been studied using atomic force microscopy (AFM). To characterize the nanostructures, the power spectral density (PSD) functions have been used. The features of the formation of periodic nanostructures are revealed depending on the incident angle of clusters and different energy per atom in clusters.

Keywords: cluster ion beam; nanostructures; potassium titanyl phosphate; atomic force microscopy; power spectral density function

Citation: Nikolaev, I.V.; Korobeishchikov, N.G. Influence of the Parameters of Cluster Ions on the Formation of Nanostructures on the KTP Surface. *Appl. Nano* **2021**, 2, 25–30. https://doi.org/10.3390/applnanao20100003

Received: 27 November 2020
Accepted: 8 January 2021
Published: 23 January 2021

Publisher's Note: MDPI stays neutral with regard to jurisdictional claims in published maps and institutional affiliations.

1. Introduction

Potassium titanyl phosphate ($KTiOPO_4$, KTP) is one of the well-known nonlinear optical materials and widely used in devices for parametric generation, high-power lasers, and electro-optical modulators with high efficiency of frequency conversion, optical waveguides in integrated optics, etc. [1,2].

It was previously shown that the monomer ion beam can form self-organized periodic nanostructures on the surface of various materials [3–6]. Such structures on surfaces attract interest for wide practical use [7]. Although gas cluster ion beams (GCIB) is commonly used for highly efficient surface smoothing, it can also be used to form nanostructures [8–12]. It should be noted that cluster ions do less damage to the subsurface layer after processing since they have a shallower impact depth and can be lower energies per atom compared to monomer ions [9–12]. As a rule, for the formation of nanostructures, cluster beams with low energies per atom in a cluster (about several eV) are used [8–11]. This work compares the efficiency of the formation of nanostructures by gas clusters with low and high energy per atom in the cluster.

2. Materials and Methods

KTP single crystals were preliminarily polished by the chemical-mechanical method. As-prepared samples are commercial single crystals of potassium titanyl phosphate made by "Siberian Monocrystal–EKSMA" Company (Novosibirsk, Russia). The KTP surface treatment by cluster ions was performed in an experimental setup described in Ref. [12]. The cluster beam was formed by adiabatic expansion of high purity argon [13]. The cluster ion beam parameters were determined using the time-of-flight technique [14]. To study the influence of the parameters of cluster ions on the processing of the KTP surface, significantly different energies per atom in cluster E/N_{mean} were selected: 12.5 eV/atom (low-energy mode) and 110 eV/atom (high-energy mode). The kinetic cluster energy E was 10 and 22 keV with the average cluster size N_{mean} = 800 and 200 atoms/cluster, respectively. In addition, we varied the incident angles of the cluster-ion beam from 0° to 70° (from the surface normal). For the correct comparison of surfaces after processing, we used the same sample.

The target surfaces were characterized using atomic force microscopy (AFM, NTegra Prima HD). The scan size of AFM measurements was 2×2 µm^2 with a resolution of 1024×1024 pixels. Additionally, we analyzed the power spectral density (PSD) functions before and after processing. The PSD function is the Fourier transform of the distribution of irregularity heights on the surface, and it is the function depending on spatial frequency ν [15–17]. We clarify here that the spatial frequency ν is a measure of how often the same height repeats per unit of distance [14]. The PSD function is helpful to determine the effective roughness σ_{eff} [12], which takes into account the lateral roughness as opposed to the convention root–mean–square (RMS) roughness R_q.

The effective roughness σ_{eff} and the PSD function are related as follows:

$$\sigma^2_{eff} = 2\pi \int_{\nu_{min}}^{\nu_{max}} PSD_{2D}(\nu)d\nu, \tag{1}$$

where $\nu_{min} = 1/L$ and $\nu_{max} = 1/\Delta x$, L—linear size of the scan area, and Δx—distance between adjacent measured points.

3. Results and Discussion

3.1. Formation of Nanostructures on the Surface of KTP Single Crystals

The maximum ion fluencies were 2×10^{16} ions/cm^2 in the low-energy mode (Figure 1) and 3×10^{15} ions/cm^2 in the high-energy mode (Figure 2). Despite the fact that the ion fluence in the high-energy mode was an order of magnitude lower, we assume that it is more correct to compare at the same etching depths because the formed nanostructures are nothing more than displaced surface material [8–11]. An aperture mask was used to estimate the etching depth and to obtain many processed areas (with different angles of incidence) on one sample. After processing, a step was formed at the "processed region–unprocessed region" boundary, the height of which was measured by AFM. The average etching depth $\langle H_{etching} \rangle$ was estimated from measurements in 9 areas of the formed step. The average etching depths were close—$\langle H_{etching} \rangle_1 = 40$ nm for low-energy mode and $\langle H_{etching} \rangle_2 = 55$ nm for high-energy mode.

Figure 1 shows 2D images and profiles of the KTP surface as-prepared and after processing by argon cluster ions at the low-energy mode ($E/N_{mean} = 12.5$ eV/atom) at the different incident angles. The lowest point of the profile corresponds to $Z = 0$ nm. A slight change in the surface profile is observed after treatment at the incident angles less than 45°, and the maximum height R_t decreases from 2.5 to 2 nm. At larger incident angles ($\geq 45°$), the maximum height R_t increases significantly (by 16–24 times compared to the original surface) due to the formation of periodic nanostructures. As can be seen in Figure 1, the highest height of ripples is observed at the incident angle of 70° (Figure 1f) but the most ordered (including laterally) ripples—at 45° and 60° (Figure 1d,e). For clarity of differences in profiles both at different angles and for different treatment modes, the maximum height on the scale (in Figures 1 and 2) is selected according to the height of the highest surface profile (Figure 1f). As it is shown in Ref. [18], such nanostructures contribute to the quality improvement of the antireflective coatings.

Figure 2 shows 2D images and profiles of the KTP surface as-prepared and after processing by argon cluster ions at the high-energy mode ($E/N_{mean} = 110$ eV/atom). First, periodic nanostructures are not observed even at incident angles of 45° and 60° (Figure 2d,e). Second, the maximum height of ripples is 10 nm at 70° (Figure 2f)—by 3 times lower than in low-energy mode. At first glance, it would seem that the high-energy treatment mode has better formation of periodic nanostructures due to high cluster energy. As seen in Figure 2, this is not the case.

We assume that this is due to the significant difference in sputtering yields of the two treatment modes—by 36 times: 100 cluster ions with $E/N_{mean} = 110$ eV/atom sputter 72 target atoms and 100 cluster ions with $E/N_{mean} = 12.5$ eV/atom sputter only 2 target atoms. Thus, a large proportion of cluster energy at the low-energy mode is spent not only on sputtering (low sputtering yield) but also on the active lateral displacement of subsurface target atoms, i.e., the formation of periodic nanostructures.

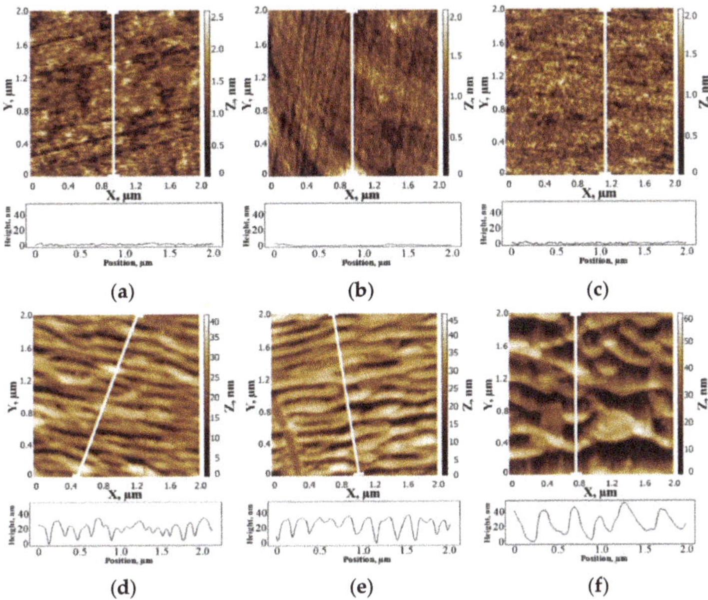

Figure 1. Atomic force microscopy (AFM) images and profiles of potassium titanyl phosphate (KTP) surface before (**a**) and after the low-energy treatment mode at different incident angles of cluster ions: (**b**) 0°, (**c**) 30°, (**d**) 45°, (**e**) 60°, and (**f**) 70°.

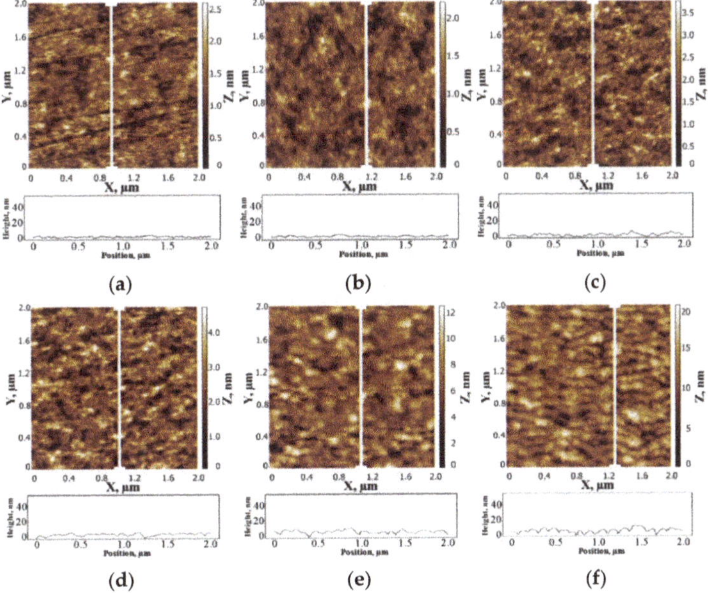

Figure 2. AFM images and profiles of KTP surface before (**a**) and after the high-energy treatment mode at different incident angles of cluster ions: (**b**) 0°, (**c**) 30°, (**d**) 45°, (**e**) 60°, and (**f**) 70°.

3.2. Power Spectral Density Functions

Figure 3 demonstrates the PSD functions of the KTP surface after processing at the different incident cluster angles for both treatment modes. In contrast to AFM images, PSD functions provide more quantitative data, especially for nanostructures. As can be seen in Figure 3a, low-energy mode smooths the KTP surface at small angles ($\leq 30°$) in the wide range of spatial frequencies ν from 1 to 8 μm^{-1} (30°) and to 15 μm^{-1} (0°). The self-organized nanostructures increase the surface roughness over the whole range of spatial frequencies. In Figure 3a, the pronounced peaks (marked by arrows) are observed. Each peak characterizes the frequency of self-organized nanostructures. It should be noted that the peak shift corresponds to the change of characteristic size (wavelength) of the nanostructure, which increases with increasing the incident angle, i.e., 167, 200, 334–500 nm for 45°, 60°, and 70°, respectively. PSD functions of high-energy mode are fundamentally different from the previous mode. First, the cluster ions with E/N_{mean} = 110 eV/atom at the incident angle of 0° better smooth the surface irregularities with the spatial frequency ν above 15 μm^{-1}, but worse smooth—with $\nu < 15$ μm^{-1}. At the incident angles of 45–70°, PSD peaks are not observed and all functions are below low-energy-functions. This corresponds to a smoother surface in whole range of spatial frequencies ν and the absence of even weakly pronounced periodic nanostructures. Thus, the nanostructures observed on AFM images (Figure 2e,f) are nonperiodic, because they have different spatial frequencies ν in accordance with the PSD-data.

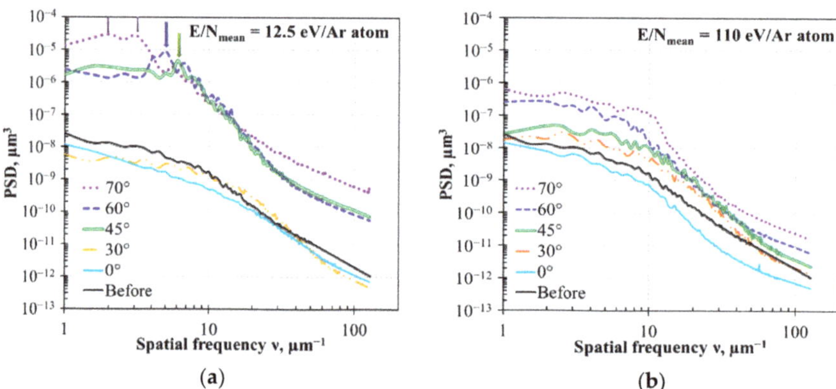

Figure 3. Power spectral density (PSD) functions of KTP surface at different incident angles of cluster ions: (**a**) after low-energy mode and (**b**) after high-energy mode.

Table 1 shows a comparison of effective roughness parameter σ_{eff} and RMS roughness parameter R_q. At both treatment modes, the surface roughness decreases to 0.28–0.30 nm at small angles ($\leq 30°$) and then significantly increases to several nanometers. It should be noted that effective roughness σ_{eff} is higher than RMS roughness R_q at low-energy mode (8–9.7%), which is most likely due to the allowance for the lateral roughness of periodic nanostructures.

Table 1. Comparison of roughness parameters.

Treatment Mode	Roughness Parameter	Before	Incident Angle of Clusters				
			0°	30°	45°	60°	70°
Low-energy mode	$\langle R_q \rangle$, nm	0.40	0.28	0.28	6.2	7.5	12.6
	$\langle \sigma_{eff} \rangle$, nm	0.43	0.28	0.30	6.8	8.1	13.8
High-energy mode	$\langle R_q \rangle$, nm	0.40	0.30	0.56	0.75	1.7	2.8
	$\langle \sigma_{eff} \rangle$, nm	0.43	0.30	0.60	0.83	1.8	2.7

4. Conclusions

The influence of fundamentally different treatment modes by the argon cluster ions on the nanostructure formation on the surface of KTP single crystals has been investigated experimentally. At normal incidence of low- and high-energy cluster ions, the surface roughness decreases to 30%. The ordered nanostructures were not observed after processing by the high-energy cluster ions (E/N_{mean} = 110 eV/atom) even at slightly higher etching depths (compared to low-energy mode) and at the small angles ($\leq$30°) of incidence of the cluster ions with E/N_{mean} = 12.5 eV/atom. It is assumed that the treatment by cluster ions with low energy leads to the formation of periodic nanostructures since the energy is insufficient for efficient sputtering of the target, but sufficient for a significant lateral displacement of subsurface target atoms. The most ordered nanostructures are formatted at the surface processing at angles of 45° and 60°, and they have a wavelength of 167 and 200 nm, respectively. One of the applications is the formation of periodic nanostructures on substrates of antireflective coatings to improve their quality.

Author Contributions: Conceptualization, I.V.N. and N.G.K.; methodology, N.G.K.; data acquisition, I.V.N. and N.G.K.; data processing, I.V.N.; original draft preparation, I.V.N.; review and editing, N.G.K.; visualization, I.V.N.; supervision, N.G.K. All authors have read and agreed to the published version of the manuscript.

Funding: This research was funded by the Ministry of Science and Higher Education of the Russian Federation, grant number FSUS-2020-0039.

Institutional Review Board Statement: Not applicable.

Informed Consent Statement: Not applicable.

Data Availability Statement: Data is contained within the article.

Acknowledgments: The authors acknowledge the NSU Shared Equipment Center "Applied Physics."

Conflicts of Interest: The authors declare no conflict of interest. The funders had no role in the design of the study; in the collection, analyses, or interpretation of data; in the writing of the manuscript; or in the decision to publish the results.

References

1. Sorokina, N.I.; Voronkova, V.I. Structure and properties of crystals in the potassium titanyl phosphate family: A review. *Crystallogr. Rep.* **2007**, *52*, 80–93. [CrossRef]
2. Mamrashev, A.; Nikolaev, N.; Antsygin, V.; Andreev, Y.; Lanskii, G.; Meshalkin, A. Optical Properties of KTP Crystals and Their Potential for Terahertz Generation. *Crystals* **2018**, *8*, 310. [CrossRef]
3. Pohl, K.; Bartelt, M.C.; de la Figuera, J.; Bartelt, N.C.; Hrbek, J.; Hwang, R.Q. Identifying the forces responsible for self-organization of nanostructures at crystal surfaces. *Nature* **1999**, *397*, 238–241. [CrossRef]
4. Frost, F.; Ziberi, B.; Höche, T.; Rauschenbach, B. The shape and ordering of self-organized nanostructures by ion sputtering. *Nucl. Instr. Meth. Phys. Res. B* **2004**, *216*, 9–19. [CrossRef]
5. El-Atwani, O.; Ortoleva, S.; Cimaroli, A.; Allain, J.P. Formation of silicon nanodots via ion beam sputtering of ultrathin gold thin film coatings on Si. *Nanoscale Res. Lett.* **2011**, *6*, 403. [CrossRef] [PubMed]
6. Ziberi, B.; Frost, F.; Höche, T.; Rauschenbach, B. Ion-induced self-organized dot and ripple patterns on Si surfaces. *Vacuum* **2006**, *81*, 155–159. [CrossRef]
7. Barth, J.V.; Costantini, G.; Kern, K. Engineering atomic and molecular nanostructures at surfaces. *Nature* **2005**, *437*, 671–679. [CrossRef] [PubMed]
8. Toyoda, N.; Mashita, T.; Yamada, I. Nano structure formation by gas cluster ion beam irradiation at oblique incidence. *Nucl. Instr. Meth. Phys. Res. B* **2005**, *232*, 212–216. [CrossRef]
9. Toyoda, N.; Tilakaratne, B.; Saalem, I.; Chu, W.-K. Cluster beams, nano-ripples, and bio applications. *Appl. Phys. Rev.* **2019**, *6*, 020901. [CrossRef]
10. Zeng, X.; Pelenovich, V.; Xing, B.; Rakhimov, R.; Zuo, W.; Tolstogouzov, A.; Liu, C.; Fu, D.; Xiao, X. Formation of nanoripples on ZnO flat substrates and nanorods by gas cluster ion. *Beilstein J. Nanotech.* **2020**, *11*, 383–390. [CrossRef] [PubMed]
11. Ieshkin, A.; Kireev, D.; Ozerova, K.; Senatulin, B. Surface ripple induced by gas cluster ion beam on copper surface at elevated temperatures. *Mater. Lett.* **2020**, *272*, 127829. [CrossRef]
12. Korobeishchikov, N.G.; Nikolaev, I.V.; Roenko, M.A. Effect of argon cluster ion beam on fused silica surface morphology. *Nucl. Instr. Meth. Phys. Res. B* **2019**, *438*, 1–5. [CrossRef]

13. Korobeishchikov, N.G.; Skovorodko, P.A.; Kalyada, V.V.; Shmakov, A.A.; Zarvin, A.E. Experimental and Numerical Study of High Intensity Argon Cluster Beams. *AIP Conf. Proc.* **2014**, *1628*, 885–892.

14. Korobeishchikov, N.G.; Kalyada, V.V.; Skovorodko, P.A.; Shmakov, A.A.; Khodakov, M.D.; Shulzhenko, G.I.; Voskoboynikov, R.V.; Zarvin, A.E. Features of formation of gas cluster ion beams. *Vacuum* **2015**, *119*, 256–263. [CrossRef]

15. Duparre, A.; Ferre-Borrull, J.; Gliech, S.; Notni, G.; Steinert, J.; Bennett, J.M. Surface characterization techniques for determining the root-mean-square roughness and power spectral densities of optical components. *Appl. Opt.* **2002**, *41*, 154–171. [CrossRef] [PubMed]

16. Persson, B.N.J.; Albohr, O.; Tartaglino, U.; Volokitin, A.I.; Tosatti, E. On the nature of surface roughness with application to contact mechanics, sealing, rubber friction and adhesion. *J. Phys. Condens. Matter* **2005**, *17*, R1–R62. [CrossRef] [PubMed]

17. Martínez, J.F.G.; Nieto-Carvajal, I.; Abad, J.; Colchero, J. Nanoscale measurement of the power spectral density of surface roughness: How to solve a difficult experimental challenge. *Nanoscale Res. Lett.* **2012**, *7*, 174. [CrossRef] [PubMed]

18. Bushunov, A.A.; Tarabrin, M.K.; Lazarev, V.A.; Karasik, V.E.; Korostelin, Y.V.; Frolov, M.P.; Skasyrsky, Y.K.; Kozlovsky, V.I. Fabrication of anti-reflective microstructures on chalcogenide crystals by femtosecond laser ablation. *Opt. Mat. Express* **2019**, *9*, 1689–1697. [CrossRef]

MDPI

St. Alban-Anlage 66

4052 Basel

Switzerland

Tel. +41 61 683 77 34

Fax +41 61 302 89 18

www.mdpi.com

Applied Nano Editorial Office

E-mail: applnano@mdpi.com

www.mdpi.com/journal/applnano